ALTERNATIVE VISIONS OF POST-WAR RECONSTRUCTION

The history of post-Second World War reconstruction has recently become an important field of research around the world; *Alternative Visions of Post-War Reconstruction* is a provocative work that questions the orthodoxies of twentieth-century design history.

This book provides a key critical statement on mid-twentieth-century urban design and city planning, focused principally upon the period from the start of the Second World War to the mid-1960s. The various figures and currents covered here represent a largely overlooked field within the history of twentieth-century urbanism.

In this period, while certain modernist practices assumed an institutional role for post-war reconstruction and flourished into the mainstream, such practices also faced opposition and criticism, leading to the production of alternative visions and strategies. Spanning from a historically informed modernism to the increasing presence of urban conservation the contributors examine these alternative approaches to the city and its architecture.

John Pendlebury is Professor of Urban Conservation at Newcastle University, UK. He has written extensively about how historic cities were planned in the twentieth century and about the role that heritage and its conservation perform in the contemporary city. His previous books are *Conservation in the Age of Consensus* and *Valuing Historic Environments*. Projects include *Town and Townscape: The Work and Life of Thomas Sharp*.

Erdem Erten teaches architectural design, and architectural history and theory classes, as Associate Professor of the Department of Architecture at Izmir Institute of Technology, Turkey. His research interests include the theorization of culture and its impact on modern architecture in the post-war period, architectural journalism and the problem of the avant-garde in architecture and romanticism.

Peter J. Larkham is Professor of Planning at Birmingham City University, UK. An urban geographer by background, he has researched and written widely on urban form and conservation, with post-war reconstruction developing as a major research interest over the past decade. His books include *The Blitz and its Legacy* and *Shapers of Urban Form*.

ALTERNATIVE VISIONS OF POST-WAR RECONSTRUCTION

Creating the modern townscape

Edited by John Pendlebury, Erdem Erten and Peter J. Larkham

LONDON AND NEW YORK

First published 2015
by Routledge
2 Park Square, Milton Park, Abingdon, Oxon OX14 4RN

and by Routledge
711 Third Avenue, New York, NY 10017

Routledge is an imprint of the Taylor & Francis Group, an informa business

Library of Congress Cataloging-in-Publication Data

Alternative visions of post-war reconstruction : creating the modern townscape / edited by John Pendlebury, Erdem Erten and Peter Larkham.
pages cm
Includes bibliographical references and index.
1. City planning—History—20th century—Social aspects. 2. Postwar reconstruction. I. Pendlebury, John R., 1963– editor of compilation. II. Erten, Erdem, editor of compilation. III. Larkham, P. J. (Peter J.), 1960– editor of compilation. IV. Sonne, Wolfgang, 1965– author. Enduring concept of civic art.
NA9040.A48 2014
711'.409045—dc23
2014006109

ISBN: 978-0-415-58734-1 (hbk)
ISBN: 978-0-415-58735-8 (pbk)
ISBN: 978-1-315-77916-4 (ebk)

Typeset in Bembo
by Apex CoVantage, LLC

Printed and bound by CPI Group (UK) Ltd, Croydon, CR0 4YY

CONTENTS

FIGURES

CONTRIBUTORS

Harriet Atkinson is Lecturer in History of Art and Design at the University of Brighton. Her monograph *The Festival of Britain: A Land and its People* was published by I.B. Tauris in 2012.

Francesca Bonfante is Associate Professor at the Department of Architecture, Built Environment and Construction Engineering ABC, Politecnico di Milano, where she teaches architectural design at the School of Civil Architecture and is also president of the Study Course in Architectural Design. She holds a PhD in architectural composition from IUAV, Venice. Her research on the relationship between specific functions and architectural experimentations for new forms of cityscape has compared the Milanese context with other European cities. Her key publications include Bonfante, *Ivrea: un cantiere per la "comunità" di Olivetti,* in *La "Città Verde"* (Boves: ArabaFenice, 2009); Bonfante & Acuto, *Sport, città e tempo libero* (Santarcangelo di Romagna: Maggioli, 2011); and Bonfante, *La costruzione del nuovo centro: Les Gratte-Ciel a Villeurbanne,* in *Anatomia di un edificio* (Napoli: CLEAN, 2012).

Nicholas Bullock is Emeritus Reader in Architectural and Planning History and a fellow of King's College, Cambridge. He teaches the history of architecture and architectural theory and the history of planning. His last book, *Building the Post-War World, Modern Architecture and Reconstruction in Britain* (London: Routledge, 2002), focused on reconstruction in Britain. His current research has extended to issues of reconstruction in Europe, particularly France and Germany, after the war.

Barnabas Calder is a Lecturer in Architecture at Liverpool University and a historian of architecture specializing in British architecture since 1945. He is currently writing a book on British Brutalism, to be published by William Heinemann, and an online complete works of Denys Lasdun, funded by the Graham Foundation and in collaboration with the RIBA British Architectural Library Special Collections.

Eamonn Canniffe leads the Architecture Research Centre and the MA in Architecture + Urbanism at the Manchester School of Architecture. He was educated in architecture at Cambridge and Harvard Universities. In 1996 he held a Rome Scholarship in the Fine Arts at the

British School at Rome. Between 1986 and 1998 he taught at the University of Manchester School of Architecture, and between 1998 and 2006 at the University of Sheffield School of Architecture. He is the author of *Urban Ethic: Design in the Contemporary City* (Routledge, 2006) and *The Politics of the Piazza: The History and Meaning of the Italian Square* (Ashgate, 2008). He is co-author (with Tom Jefferies) of *Manchester Architecture Guide* (1999) and (with Peter Blundell Jones) of *Modern Architecture through Case Studies 1945–1990* (Architectural Press, 2007). He has served for several years as Architecture Series Editor for Ashgate Publishing.

Filippo De Pieri is Assistant Professor at the Politecnico di Torino, Italy. His research focuses on the history of nineteenth- and twentieth-century European cities. His publications include *Il controllo improbabile: progetti urbani, burocrazie, decisioni in una città capitale dell'Ottocento* (Milan: FrancoAngeli, 2005) and *Storie di case: abitare l'Italia del boom* (with B. Bonomo, G. Caramellino and F. Zanfi; Rome: Donzelli, 2013).

Erdem Erten trained as an architect at Middle East Technical University in Ankara, Turkey, and completed his master's and doctoral studies in Massachusetts Institute of Technology's History Theory Criticism Program in the Department of Architecture. He teaches architectural design, and architectural history and theory classes, as Associate Professor of the Department of Architecture at Izmir Institute of Technology in Turkey. His research interests include the theorization of culture and its impact on modern architecture in the post-war period, architectural journalism, the problem of the avant-garde in architecture and romanticism.

Andrea Flores Urushima is a Researcher at the Kyoto University Center for Integrated Area Studies (CIAS) and at the Research Center for East Asian Civilizations (CRCAO, UMR CNRS) in Paris. After working as an architect-planner and studying at the Architecture and Urbanism College of the University of São Paulo in 2000, she has researched urban planning and the history of modern planning at Kyoto University since 2003.

Peter J. Larkham is Professor of Planning at Birmingham City University. An urban geographer by background, he has researched and written widely on urban form and conservation, with post-war reconstruction developing as a major research interest over the past decade. He has recently edited *The Blitz and its Legacy* (with Mark Clapson: Ashgate, 2013), written the introduction for a reprint of the classic reconstruction study *When We Build Again* (Routledge, 2013) and edited *Shapers of Urban Form* (with Michael Couzen: Routledge, 2014).

Peter L. Laurence is Assistant Professor and Director of Graduate Studies at Clemson University School of Architecture. He holds a Master of Architecture from Harvard Graduate School of Design and a PhD in architectural history and theory from the University of Pennsylvania. His research focuses on architecture culture of the 1950s, histories of architectural criticism and the field of urban design and, for the past fifteen years, the work of Jane Jacobs. His book on Jacobs is forthcoming from Penn Press, and he is at work on a reader's guide to *Death and Life of Great American Cities* for Routledge's Guides to the Great Books series.

Keith D. Lilley is Reader in Historical Geography at Queen's University Belfast. His main research interest is medieval historical geography, including the use of spatial technologies to research medieval urban form. He has also worked with Peter Larkham on post-war reconstruction, compiling an annotated bibliography of British reconstruction plans (Inch, 2001) and co-authoring papers including the study of reconstruction plan exhibitions (*Town Planning Review,* 2012).

Cristina Pallini is Senior Researcher at the Department of Architecture, Built Environment and Construction Engineering ABC, Politecnico di Milano, teaching architectural design at the School of Civil Architecture. She holds a PhD in architectural composition from IUAV, Venice. Her research on city reconstruction in the East Mediterranean has been supported by scholarships from the Aga Khan Program at Massachusetts Institute of Technology, and the Alexander S. Onassis Foundation. Her key publications include Pallini and Ferro, *Alexandria beyond the Myth* (ArabaFenice, 2009); Pallini and Recalcati, *Città Porto. Matrici, Architettura, Paesaggio* (Libraccio, 2012); and Pallini, Fiorese and Ferro, *Alessandro Christofellis: An Architectural Journey* (ArabaFenice, 2012).

John Pendlebury is Professor of Urban Conservation at Newcastle University. His research falls principally within two broad themes. First, he undertakes historically focused work, principally on how historic cities have been planned in the past, particularly in the mid-twentieth century, considering how the historic qualities of such cities were conceived and balanced with modernizing forces. Second, he undertakes empirical and conceptual work on the interface between contemporary cultural heritage policy and other policy processes. His book *Conservation in the Age of Consensus* (2009, Routledge) draws some of these themes together. His projects include the Arts and Humanities Research Council-funded *Town and Townscape: The Work and Life of Thomas Sharp,* and he has written the introduction for the reprint of Sharp's classic *Anatomy of the Village* (Routledge, 2013).

David I. Snyder holds a PhD in architectural history from Princeton University and is currently a Lecturer at Tel Aviv University in the International BA Program in the Liberal Arts. He is preparing a study examining the appearance and application of the concept of Jewish space in the shaping of the modern metropolis in central and east Europe.

Wolfgang Sonne is Professor for History and Theory of Architecture at the Faculty of Architecture and Civil Engineering at the Technische Universität Dortmund and deputy director of the Deutsches Institut für Stadtbaukunst (German Institute for Civic Art). He studied art history and archaeology in Munich, Paris and Berlin and holds a PhD from the Eidgenössische Technische Hochschule (ETH) in Zurich. He has previously taught at the ETH Zurich and at the University of Strathclyde in Glasgow. His publications include *Representing the State: Capital City Planning in the Early Twentieth Century* (Prestel, 2003) and *Urbanität und Dichte im Städtebau des 20. Jahrhunderts* (DOM Publishers, 2013).

ACKNOWLEDGEMENTS

The genesis of this book was a conference, *Visual Planning and Urbanism,* held in Newcastle upon Tyne in 2007, where approximately half of the papers included in this volume first saw the light of day. The conference was supported as part of the Arts and Humanities Research Council-funded project *Town and Townscape: The Work and Life of Thomas Sharp.* This coincided with Erdem Erten spending three months in Newcastle as British Academy/Economic and Social Research Council Visiting Fellow for South Asia/Middle East working on *The Development of Townscape as an Urban Design and Conservation Pedagogy: The Architectural Review and the Work of Thomas Sharp.* We would therefore like to acknowledge the contributions of the AHRC, British Academy and ESRC for making this book possible. More information on the *Town and Townscape* project, and Thomas Sharp, can be found at http://catless.ncl.ac.uk/sharp. We would also like to thank Laura Fernandez Gonzalez for her tremendous efforts in the early stages of this work, Adam Sharr for useful comments on Chapter 1 and Liz Brooks for her careful copy-editing as this volume finally came together. Finally, we would like to thank all at Routledge for their patience and support for this book.

PART I

Introduction

1

ON "ALTERNATIVE VISIONS"

Erdem Erten, John Pendlebury and Peter J. Larkham

Introduction

The twenty-first century is often characterized as "the urban century", as for the first time more than half of humanity – 52 per cent by 2011 – is living in urban areas.[1] However, the foundations for the urban century were very firmly laid during the twentieth century, as processes of rapid urbanization, experienced in some countries in the nineteenth century, became established as a global phenomenon. This brought an unparalleled increase in the outward expansion, and in the building, demolition and rebuilding, of cities. Approaches and attitudes to managing this rapidly developing environment also underwent remarkable changes during the course of the twentieth century, with accompanying major changes in and redefinitions of the professional disciplines that shape the city. Thus, whilst the planning and design of towns and cities remains an activity that transcends an urban history of several thousand years, it was during the twentieth century that a profession of town planning emerged, having split from the older established professions of architecture and surveying, and incorporating other developing social sciences, such as sociology.

There is a historiographic consensus that periodizes planning into roughly five distinct periods, typically the pre-industrial revolution (that is before planning per se, although there were many examples of conscious urban layouts in the classical, medieval and Renaissance periods), from the industrial revolution to 1900, from 1900 to the Second World War, from the Second World War to the early 1970s oil crisis, and from then to the present.[2] Within this periodization, the most dramatic turning point is the Second World War, which brought an unprecedented scale of destruction, the impact of which was most manifest in cities. It was a catalyst for change even in countries suffering relatively little direct war damage (for example Australia) and in non-combatant countries (such as those of South America). The principal aim of this book is to provide a critical account of mid-twentieth-century urban design and city planning, focused principally upon the period between the start of the Second World War and the early 1970s, when the term "urban design"[3] began to be adopted and to incorporate themes from an earlier urbanism. Surprisingly, this important period is understudied. This is perhaps a controversial assertion as much has been written since Diefendorf's pioneering edited volume on European bombed cities.[4] This work has included studies of individual planners and their plans,[5] of processes of planning and administration,[6] of the politics of people and processes[7] and of the communication of planning ideas.[8] Newer perspectives to emerge include more consideration of those affected by the replanning and rebuilding[9] or of the link between reconstruction and culture.[10] Comparative work remains relatively scarce,[11] and, to our knowledge, little has been published in English on reconstruction in countries such as the former USSR, China and those in North Africa.

Precursors: the late nineteenth to early twentieth century

Early-twentieth-century urban visions were largely characterized by a demand for the radical transformation of the nineteenth-century post-industrial revolution city, which was often seen as something to be obliterated instead of rehabilitated. Such cities were thought to be no longer comprehensible as a totality and were frequently described as unhealthy bodies requiring radical surgery, by means almost completely linked to advances in transport and other infrastructure. Haussmann's Paris boulevards are a seminal early example. The end of the nineteenth and beginning of the twentieth centuries brought more radical, largely unbuilt reimaginings, such as those of the Italian Futurists or the ideas for linear cities proposed alternatively in Spain and, later, the Soviet Union. Volker M. Welter has illustrated how influential writers such as Lewis Mumford, Sigfried Giedion and Arthur Korn self-consciously analysed the evolution of cities as part of a platform for arguing for radical urban change.[12] The faith in such restructuring and the rise of planning largely stemmed from belief in the progressive ideals of the Enlightenment and the ensuing modernism, and from the conviction that the urban environment and the social life it housed could – and indeed should – be planned. Planning was invoked to harness the apparently unstoppable growth of urban agglomerations, not only metropolises but also small towns, in response to perceptions of the declining quality of life in the city and the urban, and the suburbanization of the countryside through the impact of sprawl in various ways, such as ribbon development and planned suburbs.

In addition to the separation of engineering from architecture, and its institutionalization in the late nineteenth century, the turn of the twentieth century coincided with the further advance of specialization within the disciplinary structures that shape the built environment. The most famous British vision of planning as reform, Howard's Garden City, always had a strong social dimension, and this was true for other important pioneers of the profession, such as Patrick Geddes. However, in the demands for the recognition of planning as a separate profession and academic discipline, with its own institutional bodies, the discipline came to be focused on the physical planning of the built environment and the macroform of the city. This is typified by the foundation of the UK's Town Planning Institute and the beginnings of the separation of planning activities from the Royal Institute of British Architects.[13] Framed in this manner, planning was seen more as a largely *beaux-arts* architecture-driven art, with its social dimensions marginalized. This began, particularly after the First World War, to give way to a more structured, 'scientific' attitude. Throughout the 1930s, as the canonical historiography of planning narrates,[14] the dominant responses of architecture and planning were CIAM's (Congrès Internationaux d'Architecture Moderne's) functional city and versions of the Garden City, together with urban utopias that turned away from the complexities that paralysed the metropolis, as well as de-urbanist schemes that looked towards its dissolution in the countryside. The visions of the principal members of CIAM, and Le Corbusier in particular, came to hold a particularly strong grip on the modernist-focused histories of the mid-twentieth century.[15] CIAM did, of course, partly exist precisely for the purpose of "creating a unified sense of what is now usually known as the Modern Movement in architecture".[16] It was trying to create a party line, even if its chroniclers now acknowledge that CIAM overstated its claim to be the representative organization of modern architecture[17] and that the existence of a party line was, in practice, illusory.[18]

The post-Second World War period

From one point of view this period is seen as an interlude between the rise of CIAM and the ultimate collapse of modernist narratives and the rise of postmodernism, but from another it represents an extended but productive crisis that led to the emergence of various operative responses based on a critical re-evaluation of modernism. As recent research (some already cited) has indicated, this period has a complex and rich history that cannot be reduced to the ascendancy of particular modernist urban visions such as the Garden City and its suburban variants, the zoned offshoots of Athens Charter urbanism inspired by Le Corbusier's obsession with mobility and speed, or the sprawling expanses of Frank Lloyd Wright's Broadacre City in the North American context dispersed for the nuclear age. In this period, while certain modernist practices – impelled at least in part by the exigencies of wartime mass production[19] – assumed an institutional role for post-war reconstruction and flourished into the mainstream, such practices also faced opposition and criticism from within the disciplines. Although in many places the professional planners, architects and surveyors were seen as the experts, their visions also began to generate adverse public responses. These trends led to the production of alternative visions and strategies, sometimes arising from members of the public, press campaigns or pressure groups, as well as from re-evaluations within the professions themselves. Spanning from a historically informed modernism to the increasing influence of urban conservation, this book aims to cover a range of these alternative approaches to the post-war city, its redevelopment and its architecture. The various individuals, organizations and trends of thought and practice investigated by the contributors represent an under-researched field within the history of twentieth-century urbanism despite the attention paid in recent years to a small number of these figures, such as Thomas Sharp, or the idea of Townscape that developed around the *Architectural Review*'s editorial campaigns.[20] This volume seeks to extend consideration to some of the wider issues within which contemporary design and planning operated, thus (for example) extending Whyte's related collection of papers on aspects of design education in this period.[21] It is also true that some of the familiar key players have been treated only very partially by the existing literature, and we shed new light on some of their individual and collective activities. Some of these alternative visions have now become part and parcel of mainstream urban design practices after a lengthy process of absorption and appropriation, although their histories are still being written.

Alternative visions

The alternative visions cited in this book belong to thinkers and practitioners who were acutely aware, and critical, of the problems and deficiencies of planning and of its importance for post-war reconstruction, and who, whilst often "modernists", remained largely outside the mainstream of modernist planning and architecture. We argue that these protagonists, consciously or unconsciously, effectively laid the foundations of what came to be termed "urban design". This book therefore focuses on the birth of urban design, via the critique of urban planning and the revival and reinterpretation of themes of an earlier urbanism when planning and architecture were more unified.

The period from the end of the Second World War until the late 1960s or early 1970s might be thought of as the heyday of physical planning. In simple terms, whilst there is no clear East–West dichotomy, in much of the 'West' (or the post-Second World War Western bloc) the welfare state – which became the dominant governmental model – relied heavily on comprehensive planning, and in the 'East' (the Eastern bloc) the state assumed more or less complete control. However, physical planning was largely abandoned in favour of policy planning towards the end of the 1970s. This was accompanied by the gradual emergence of urban design as a semi-autonomous field of activity connecting planning and architecture. Although this transition from the abandonment of physical planning to the turn towards policy and advocacy planning catalysed the emergence of urban design as a distinct epistemology and field of practice, it is not the sole explanation. Chronologically, and perhaps also functionally, the decline of civic design and its inherent visual dimension is closely associated with the rise of physical planning – technocentric and data-driven and, in the post-war replanning context in Britain, for example, strongly linked to the dominant influence of the 'Planning Technique' section of the new Ministry of Town Planning. The alternative visions of this book were promoting a visual approach to planning and at the same time sowing the seeds that grew to be urban design. Thus, we argue, a historical recontextualization of the period demands a re-reading of the resistance to orthodox modernist institutions of governance, education and professionalism, and the associated development industry, to the split between architecture and planning, and the role of this resistance in the birth of urban design.

Orthodox modernist planning undermined certain qualities of the environment that were left out of the representational devices of physical planning, as in the post-war period planning simultaneously became a more technocratic and expertise-demanding endeavour within which the principal access point for the public into the future of a city was its visual representations. But these became noticeably fewer. The activity of planning itself became notable for the proliferation of technocentric representations which were largely map-based or diagram-based, denoting land-use allocations with set colours and the like, addressing a very limited and professionally educated audience. This became characterized as the "planner's gaze", the view from above, focusing on the data underlying the land-use allocation decisions.[22] Therefore, while the purging of the visual from the field reduced planning's communicative possibilities, at the same time it led to a reduction in understanding of the built environment's qualities. One of the reasons for the reaction against modernist planning in the post-war period was the increasing detachment of planning from the public: its communication was perceived as poorer, and its outputs were less readily comprehensible.[23]

The disciplinary split within the field of urbanization brought a reshuffling of the cards, that is a reorganization of the professional field under a new specialization. Elements purged from the field of planning, with reference to both scale and the visual, returned in the burgeoning field of urban design, although the term "urban design" itself was adopted and recognized later, towards the end of the 1950s. In its early days, that is immediately after the war, this comeback was largely resisted in orthodox circles as a conservative reaction to modernism, although the early protagonists of urban design, who made explicit their connections to visual planning, did not necessarily see themselves in these terms.

Whilst the term "urban design" may be relatively modern it has been referred to as "an activity with ancient roots that has been recently rediscovered and reinvigorated",[24] one that deals largely with urban form and public space and has developed a growing body of theoretical writing. Urban design readers attribute canonical status to books by Kevin Lynch, Gordon Cullen and Jane Jacobs within this rebirth as early pioneers in urban design discourse – all of these were first published in the early 1960s. As this book aims to make clearer, the abovementioned work came to fruition only after a longer period of distillation via architectural journalism, academia and practical planning efforts. Historians like to distinguish between civic design and urban design by portraying civic design as "highly artistic and physical", concerned more with monuments and open spaces, typified by the City Beautiful movement.[25] Although one can draw temporal and conceptual distinctions between them, Wolfgang Sonne in this volume argues that urban design and civic design are of the same genealogy and have more in common than is normally accepted. Indeed, clear bridges between the two movements can be identified in some contexts. For example, in the UK Arthur Trystan Edwards self-consciously worked in a civic design tradition[26] and influenced Thomas Sharp, who was, in turn, one of the leaders of the Townscape discourse that fed into the nascent subject of urban design. Urban design has subsequently flourished and grown into something else, but in its emergence, in the period covered by this book, it owed much to civic design.

This book intends to enrich the history of post-war urbanization by laying bare its complexities that remain outside the period's canonical historiography. Considering the idea of "alternative visions", two meanings of the word 'vision' and their coming together within the context of urbanization are of particular interest here: first, "the action of seeing with the bodily eye the exercise of the ordinary faculty of sight, or the faculty itself"; and, second, the "mental concept of a distinct or vivid kind; an object of mental contemplation, esp. of an attractive or fantastic character; a highly imaginative scheme or anticipation".[27] In other words, the chapters of the book explore a broad range of exemplars of urbanization and post-war reconstruction, in different geographical (and hence socio-political and economic) contexts, which have an evident highly visual emphasis. This is part of the reason why the book is subtitled "creating the modern townscape". Recent research has shown that the development of Townscape frames the post-war reinterpretation of civic design and the resurging interest in visual planning within a new historical context, in reaction to the disciplinary evolution of planning as a separate discipline from which the visual was purged as anathema, and to the apparent disconnection between modern architecture's object-centred focus and the urban fabric.[28] Townscape is part of the earliest efforts at conceptualizing in a simple way how contemporary intervention should come to terms with the existing urban fabric, that is an earlier development of the contextual sensitivity that underlies urban design today. In other words, Townscape is proto-urban design that harbours interest in historical continuity, the adaptability of urban heritage to present needs, the questioning of the impact of larger-scale planning ventures, to the needs of the pedestrian and the aesthetics of the city.

Alternative visions: the organizational framework

The book is organized into four parts. The introductory section comprises this chapter and Wolfgang Sonne's chapter 'The Enduring Concept of Civic Art', which sets out an important platform for the book. Using a wide range of examples from Europe and North America, Sonne demonstrates the continuation of an important tradition of urban design which can be linked to ideas of civic art and design early in the twentieth century and which persisted throughout the time frame covered by this book and can ultimately be linked to nascent postmodern approaches. He shows how this was a practical and pragmatic tradition that often utilized traditional forms in its urban interventions tested in the complexities of urban life and in turn was critical of the functional simplifications of the CIAM agenda. As referred to above, a figure such as Trystan Edwards is a good example of someone who bridged between early-twentieth-century civic design, was explicitly critical of Le Corbusier, but linked to the practical British modern urbanism embodied within the *Architectural Review*-promoted approach to Townscape or the planning of Thomas Sharp.

The second part, 'Imagined Townscapes', addresses the development of alternative visions for the city and the respective historical contextual backgrounds for these visions, which have not necessarily been implemented into built results but have mainly served theoretical or pedagogical purposes. The contributions by each author in this section point to a complex process of negotiation ensued by the dynamics of rapid change in different urban contexts, especially in the realm of urban design thinking, the results of which became manifest in the longer run.

The idea of Townscape is usually discussed in terms of its use as a visual planning methodology.[29] However, Erdem Erten's focus in chapter 3 is upon its cultural role and the use of Townscape and subsequent *Architectural Review* campaigns in turn as part of a project of cultural continuity. This was led by proprietor and editor Hubert de Cronin Hastings but was a consistent *Architectural Review* policy over a forty-year period under different editors. It was an approach to urban design that sought to reconcile modernity with tradition, but also with an ideological underpinning, equating the picturesque with the idea of (British) liberal democracy.

In post-war Italy Italian architects sought to balance modernist ideals and the impacts of accelerating modernization with pressures for the conservation of the historic city. Filippo De Pieri argues that Italian architects of the period were pushed to adapt the analytical tools of modern planning to the preservation of historical cities as they dealt with the problem of reconciling the historical with the modern. He focuses on two particularly influential plans of the period, Giovanni Astengo's unimplemented plan for Assisi and Giancarlo De Carlo's plan for Urbino, which enjoyed even greater transmission through its publication in English. Although these plans do not represent Italian planning culture as a whole, they provide significant examples of the way in which the historic city was analysed and represented by Italian planners after the Second World War, posing challenges to which planning tools grounded in CIAM and post-CIAM traditions did not appear entirely suited. "Both plans were characterized by methodological eclecticism, an approach possibly more indebted to European traditions of visual or picturesque planning than their authors were willing to admit", argues De Pieri.

While Erten investigates how the idea of Townscape has been instrumentalized in pushing forward a cultural policy embedded in architectural journalism to defend cultural continuity, Harriet Atkinson contends that "a new picturesque" conceived during the Festival of Britain (1951) was intended to change the post-war British landscape in an attempt to mediate between nostalgia for a pre-industrial landscape and the present. The organizers decided that the eight major exhibitions of the Festival (of which the South Bank was the centrepiece) would be united by the theme 'the land and the people of Britain'. Atkinson describes the rich connections between the promoters of Townscape and the South Bank Festival design team, exemplified by the Festival's director of architecture, Hugh Casson, and the picturesque principles used in the planning of the South Bank site; perhaps this was the practical exemplification of Townscape. She goes on to describe how picturesque principles underpinned the design of subsequent housing developments, including the Lansbury Estate (the Festival's Live Architecture Exhibition) and subsequently at Harlow New Town.

Andrea Flores Urushima's chapter introduces Nishiyama Uzō's scepticism towards the statistics-based biases and formalist idealism of modernist planning, and his search to accommodate the ways of life of ordinary people within urbanization and urban theory by means of analyses on housing in Japan. An architect and planner who was well informed about European modernist currents, and the son of an industrialist who witnessed the Japanese industrial revolution and its impact on people's lives, Nishiyama advocated a critical re-evaluation of international modernism against its dogmatic and ubiquitous implementation and the false pretences of a nationalist historicism. In order to validate his position Nishiyama searched for a language that would visually communicate the possible futures of a plan, good or bad, inviting citizen participation in the process of city making.

Part III, 'Townscapes in practice', addresses how, in several different cultural and geographical contexts, alternative visions were put into practice. It focuses on post-war visions, plans and theories, as well as the built environments that resulted, and explores some of the conflicts emerging from these processes. Clearly, these chapters depict only a small and unrepresentative sample of the alternative visions and townscapes of the post-war years, yet they act as a necessary grounding for the ideas explored throughout the book.

Peter J. Larkham and Keith D. Lilley explore the concept of scenography, most particularly as used in Sharp's English reconstruction plans. Although it was undermined by early modernist histories of architecture, scenography – defined as a three-dimensional view of urban space – has played a crucial role in the conceptualization and communication of ideas in urban planning and design. Such 3D visualization was very common in the advisory reconstruction plans of the period, the best-resourced of which used the best architectural illustrators, model-makers and related resources. The lay public responded well: this visual communication of complex ideas for urban futures was popular and effective, as Larkham and Lilley show for the large number of public planning exhibitions.[30] However, it may have simultaneously created a problem of false expectations of certainty and delivery, although the accompanying text of many plans explicitly noted the intentional vagueness of architectural form and detail in the future visions. Scenographic representation declined quickly with the new generation of Development Plans specified in the 1947 Town and Country Planning Act.

Sharp's key role is developed further by John Pendlebury. He was a most prolific plan-producer, and although most of Sharp's plans were for smaller or historic towns, Pendlebury shows that his ideas should not be limited to these contexts. His principles – the basis for his sometimes radical new visions – lay in his critique of Garden Cities, the loss of urbanity and the merging of Town and Country. Townscape, like landscape, could have picturesque qualities. Yet he was modernist in his treatment of the past: it was inspiration, not straightjacket. Plans for Durham, Exeter and Oxford had lengthy preambles, identifying the historic character or *genius loci*. This was, especially for Oxford, essentially a Townscape analysis and manifesto. Given this approach Sharp could propose clearing away part of bombed Exeter and creating new cathedral views in the process, rather than replicating it. He thus was brave enough to avoid obvious, popular solutions and to champion instead the occasional drastic intervention, such as his relief road for inner Oxford, Merton Mall. So Sharp's alternative visions were simultaneously bold and locally focused, locally sensitive and strategic, and his concept of Townscape was for widespread application and for the creation of high-quality urban places.

Francesca Bonfante and Cristina Pallini discuss historic townscapes in reconstruction plans in the context of Italian architectural attitudes. Here, the issue of urbanism and the rich Italian urban heritage led to markedly different approaches to other national contexts. Italian architects developed a typomorphological approach to analysing and understanding context and *genius loci*. They were aware of the social role of architecture and replanning, not least because of the large-scale population movement from the Italian South to the North. A number of key individuals and plans moved the debate substantially, although pre-war activity and a 1942 planning act were important precursors. Proposals for Milan and Turin were based on detailed historical study, a reappraisal of urban scale and function, and appreciation of the place of a city in its wider region. Outwardly modernist, they were in practice subtle in concept. In the historic centres of cities such as Genoa, proposals likewise challenged contemporary aesthetics in juxtaposing old and new construction. These plans and projects were instrumental in re-evaluating architectural modernism and identifying its impracticalities, especially when faced with an urban heritage of world-class significance as well as the magnitude of the post-war reconstruction task.

Finally in this section, David Snyder uses the example of the reconstruction of devastated Warsaw to examine rhetoric and politics in the communication of ideas, especially of the place of architectural modernism in the reconstruction and replanning of a major historic city. The catastrophe of destruction, the splendour of what had been lost and the possibilities of the future were communicated within Poland and much more widely through photographic images in publications (especially the magazine *Stolica*) and exhibitions. These were a means not only of communicating but also of building – or rebuilding, or redefining – culture and identity. A uniform collective memory was being sought. That the historic Old Town (Stare Miasto) was replicated is widely known, but the nature of authenticity was here established through the photographic representations. Once the Stare Miasto had been largely rebuilt, by 1953, the use of imagery changed, especially in the new City Museum. The 'before' images were the photographic narrative of the past and its destruction. The 'after' was now the rebuilt urban landscape. That landscape was also not just the reconstruction of the historic centre but the emerging 'new metropolis'. Here, then, the alternative visions are inherent in the deliberate presentation of a political message, reinterpretation of the past and promotion of a radical urban future (outside the Stare Miasto at least). This is both continuity and contradiction.

The final part, 'Townscapes in opposition', focuses on histories of opposition and conflict in relation to the mainstream practices of urban design and planning, and on the transformative impact of such opposition on urban design and planning discourse, as well as the resulting reconstruction efforts.

In Gaston Bardet's fierce attack against the Athens Charter, and the publication history of the French periodical *Urbanisme,* Nicholas Bullock points to a widespread opposition to the codification of Corbusian urbanism into legislation in France. He demonstrates an important division within French architectural journalism: journals such as *Architecture Française* and *Urbanisme* demanded an accommodation with older traditions of French urbanism and a recognition of conservation, and another camp, including *L'Architecture d'Aujourd'hui* and *Techniques et Architecture,* expected French city centres to emerge 'modernized' from the ruins of war and industrial decay. After the war *Urbanisme* continued to serve as a melting-pot of urbanism via its diverse editorial composition, until this was threatened as a result of the French administration's sympathy to CIAM principles for reconstruction. In its 1956 review the journal demonstrated that a variety of approaches to reconstruction coexisted, such as those that advocated integration into the historical urban fabric that had survived the war, and those that remained faithful to modernism in creating entire quarters from scratch. Bullock's contribution demonstrates that reconstruction was not simply concerned with the physically urban but also with the revitalization of culture.

Barnabas Calder further develops the theme of coexisting concepts and argues that the advocates of New Brutalism, some of whom were also members of Team 10, shared many of the concerns embodied in Townscape. He questions the representation in previous histories of an opposition between New Brutalism and Townscape, an approach that has perpetuated a hagiographic tendency which privileges the post-CIAM neo-avant-garde, especially the work of the Smithsons. Applying recent archival research, Calder demonstrates that the inter-generational breach which characterizes British post-war architectural historiography was largely manufactured as a result of the neo-avant-garde's self-characterization in search for power within the professional field.

Peter Laurence's research on Jane Jacobs reveals that together with the editors at *Architectural Forum* she was promoting an agenda for a way out of the decline of city centres in the US in collaboration with her English counterparts who were running the campaign for Townscape, a joint effort which would also define the emerging field of urban design. Laurence argues that Jacobs, Douglas Haskell and others at *Architectural Forum* were quick to see Townscape's critical dimension regarding the utopian, unworkable and joyless character of anti-city visions. In Townscape theory Jacobs found support for her own work, which valued concentration, density, centralization, pedestrianization, variety of demand, multiple use and multiplicity of choice above functional zoning and car-dependent suburban sprawl, which, she argued, gradually led to the decay of urban centres. Those interested in a similar understanding of urbanism, such as Ian Nairn, Christopher Tunnard, Ian McHarg and Kevin Lynch, would soon come together under the umbrella of the Rockefeller Foundation's grant scheme to influence the field of landscape and urban design.

Eamonn Canniffe reframes Italy's post-war reconstruction efforts as the stage of a debate where the limitations of the universal models favoured by the older generation of modernists and the intuitive approach of Townscape were critiqued by the adherents of Team 10 in an attempt to overcome such limitations by appealing to anthropology. The Italian example is particularly fascinating since this critique was produced at a time when the avant-gardist rejection of history was brushed aside in favour of historical models, and the cultural value of historical environments received renewed interest, together with the evaluation of such environments' potential for tourism.

Coda

In this volume we seek to extend knowledge of a short but key period in the development of planning thought and practice, and in the shape of the towns and cities. Although many of the numerous plans produced were short-lived or not implemented, these emerging ways of thinking, and the places that did result from them, formed the towns and cities in which so many of us live and work today. In drawing together perspectives on this period from a range of countries most affected by the destruction of the Second World War we learn more about the origin and implementation of these alternative visions, and about the individuals and circumstances behind them.

There is a wider research agenda which we aim to provoke and promote, and believe deserves more attention. There remain countries, let alone individual places and plans, about which, as far as we are aware, little systematic research has been undertaken. Likewise, individuals, groups and public and private institutions (local or global) – which are responsible for generating ideas, making and implementing plans and overall reshaping alternative visions into new professional agendas – are worth further exploration in a historical context, as the same processes are occurring today. New perspectives on these planning histories also emerge, including the still little-studied planning histories of those affected by planning, whose voices are often little heard;[31] a reconsideration of effective communication and participation; and the wider impacts of planning on society as a whole. The short period of the 1940s and 1950s still has much to teach us.

Notes

1. World Bank, Urban Development, www.data.worldbank.org/topic/urban-development, accessed 3 January 2013.
2. Nigel Taylor, *Urban Planning Theory since 1945,* London: SAGE, 1998.
3. According to Eric Mumford the term "urban design" was first adopted in the US following discussions at Harvard from the 1950s and the commencement of a programme of study in 1960. E. Mumford, *Defining Urban Design: CIAM Architects and the Formation of a Discipline, 1937–69,* New Haven: Yale University Press, 2009.
4. J.M. Diefendorf (ed.), *The Rebuilding of Europe's Bombed Cities,* London: Macmillan, 1990.
5. For example P.N. Jones, '". . . a fairer and nobler City" – Lutyens and Abercrombie's plan for the City of Hull 1945', *Planning Perspectives,* 1998, vol. 13, pp. 301–316; P. Scrivano, 'The elusive polemics of theory and practice: Giovanni Astengo, Giorgio Rigotti and the post-war debate over the plan for Turin', *Planning Perspectives,* 2000, vol. 15, pp. 3–24; L. Campbell, 'Paper dream city/modern monument: Donald Gibson and Coventry', in I.B. Whyte (ed.), *Man-Made Future: Planning, Education and Design in Mid-Twentieth-Century Britain,* London: Routledge, 2007.
6. For example J.M. Diefendorf, 'Reconstruction and building law in post-war Germany', *Planning Perspectives,* 1986, vol. 1, pp. 107–129; C. Flinn, '"The city of our dreams"? The political and economic realities of rebuilding Britain's blitzed cites, 1945–54', *Twentieth Century British History,* 2012, vol. 23, pp. 221–245.
7. For example N. Tiratsoo, *Reconstruction, Affluence and Labour Politics: Coventry 1945–60,* London: Routledge, 1990; S. Essex and M. Brayshay, 'Vision, vested interest and pragmatism: who re-made Britain's blitzed cities?', *Planning Perspectives,* 2007, vol. 22, pp. 417–442.
8. For example P.J. Larkham and K.D. Lilley, 'Exhibiting the city: planning ideas and public involvement in wartime and early post-war Britain', *Town Planning Review,* 2012, vol. 83, pp. 647–668.
9. For example K.D. Lilley, 'Conceptions and perceptions of urban futures in early post-war Britain: some everyday experiences of the rebuilding of Coventry, 1944–62', in I.B. Whyte (ed.), *Man-Made Future: Planning, Education and Design in Mid-Twentieth-Century Britain,* London: Routledge, 2007; D. Adams, 'Everyday experiences of the modern city: remembering the post-war reconstruction of Birmingham', *Planning Perspectives,* 2011, vol. 26, pp. 237–261.
10. L. Mellor, *Reading the Ruins: Modernism, Bombsites and British Culture,* Cambridge: Cambridge University Press, 2011.
11. It is relatively rare to compare different locations within one country: see J. Hasegawa, *Replanning the Blitzed City Centre,* Buckingham: Open University Press, 1992; C. Hein, J.M. Diefendorf and Y. Ishida (eds), *Rebuilding Urban Japan,* Basingstoke: Palgrave Macmillan, 2003. Comparison between countries is still rarer: see J.L. Nasr, *Reconstructing or Constructing Cities? Stability and Change in Urban Form in Post-World War II France and Germany,* unpublished PhD dissertation, Philadelphia, University of Pennsylvania, 1997; N. Tiratsoo, J. Hasegawa, T. Mason and T. Matsumura, *Urban Reconstruction in Britain and Japan, 1945–1955: Dreams, Plans and Realities,* Luton: University of Luton Press, 2002.
12. V.M. Welter, 'Everywhere at any time: post-Second World War genealogies of the city of the future', in I.B. Whyte (ed.), *Man-Made Future: Planning, Education and Design in Mid-Twentieth-Century Britain,* London: Routledge, 2007.
13. G.E. Cherry, *A History of British Town Planning,* Leighton Buzzard: Hill, 1974.
14. See for instance P. Hall, *Cities of Tomorrow: An Intellectual History of Urban Planning and Design in the Twentieth Century,* Oxford: Blackwell, 1988; or R. Fishman, *Urban Utopias in the Twentieth Century,* Cambridge, MA: MIT Press, 1982.

15. On the origins and influence of modernism and the place of CIAM, see J.R. Gold, *The Experience of Modernism,* London: Spon, 1997; J.R. Gold, *The Practice of Modernism,* London: Spon, 2007.
16. E. Mumford, *The CIAM Discourse on Urbanism, 1928–1960,* Cambridge, MA: MIT Press, 2000, p. 1.
17. Ibid.
18. S.W. Goldhagen, 'Coda: reconceptualizing the modern', in S.W. Goldhagen and R. Legault (eds), *Anxious Modernisms: Experimentation in Postwar Architectural Culture,* Cambridge, MA: MIT Press, 2000.
19. J.-L. Cohen, *Architecture in Uniform: Designing and Building for the Second World War,* Montreal: Canadian Centre for Architecture, 2010.
20. J. Pendlebury, 'The urbanism of Thomas Sharp', *Planning Perspectives,* 2009, vol. 24, pp. 3–28; E. Erten, 'Thomas Sharp's collaboration with H. de C. Hastings: the formulation of townscape as urban design pedagogy', *Planning Perspectives,* 2009, vol. 24, pp. 29–49; and E. Erten, *Shaping 'the Second-Half Century': The Architectural Review 1947–1971,* unpublished PhD dissertation, Massachusetts Institute of Technology, Cambridge, MA, 2004. For a recent compendium of research on Townscape see Mathew Aitchison (guest ed.), 'Townscape revisited', special issue of *Journal of Architecture,* 2012, vol. 17, no. 5. For other figures see S.W. Goldhagen and R. Legault (eds), *Anxious Modernisms: Experimentation in Postwar Architectural Culture,* Cambridge MA: MIT Press, 2000.
21. I.B. Whyte (ed.), *Man-Made Future: Planning, Education and Design in Mid-Twentieth-Century Britain,* London: Routledge, 2007.
22. See the critique in M. de Certeau, 'Walking in the city', in *The Practice of Everyday Life,* trans. S. Randall, Berkeley: University of California Press, 1984.
23. Jane Jacobs's journalistic work at the Architectural Forum and the *Architectural Review*'s post-war campaigns frequently take issue with this communication divide between experts and citizens. See also J. Jacobs, *The Death and Life of Great American Cities,* New York: Random House, 1961; and E. Erten, 'Postwar visions of apocalypse and architectural culture: the *Architectural Review*'s turn to ecology', *The Design Journal,* 2008, vol. 11, no. 3, pp. 269–285; or Erten, *Shaping,* op. cit.
24. M. Carmona and S. Tiesdell, 'Introduction', in M. Carmona and S. Tiesdell (eds), *Urban Design Reader,* Oxford: Architectural Press, 2007, p. 1.
25. See especially Mumford, *Defining Urban Design,* op. cit.; Carmona and Tiesdell, 'Introduction', op. cit.; and M. Larice and E. Macdonald, *The Urban Design Reader,* London: Routledge, 2007.
26. See A. Trystan Edwards, *Good and Bad Manners in Architecture* (revised edn), London: Tiranti, 1946.
27. Definitions from the online *Oxford English Dictionary,* www.oed.com, accessed 14 January 2013.
28. See for example special issue of *Journal of Architecture,* 2012, vol. 17.
29. See for example M. Aitchison, 'Townscape: scope, scale and extent', *Journal of Architecture,* 2012, vol. 17, pp. 621–642.
30. Larkham and Lilley, 'Exhibiting the city', op. cit.
31. L. Sandercock, *Making the Invisible Visible: A Multicultural Planning History,* Berkeley: University of California Press, 1998.

2

THE ENDURING CONCEPT OF CIVIC ART

Wolfgang Sonne

The concept of Civic Art (*Stadtbaukunst*) formed the core of modern urban development theory in its attempt to break away from the practice and concepts of the urban expansion of the nineteenth century. The term 'art' related to two things: first, it referred to the aesthetic–formal aspect of the city, the artistic–design perspective, and it is here that the promoters of Civic Art (with its inclusion of aesthetics) consciously distanced themselves from the technical and engineering focus that had characterized urban development in the nineteenth century. Second, it indicated the importance of considering the aesthetic–formal in relation to other aspects of the city such as society, politics, economy, technology, culture and nature, as well as to the other tasks of urban development such as solving problems of housing, transport and recreation. Thus, Civic Art in this period was not just about design and art; it also considered their interplay with other vital factors for the city. In this respect, Civic Art advocated a comprehensive and sophisticated framework – one which was largely abandoned by subsequent avant-garde movements and reduced to only a few factors. The result of this was the purely transport-related, sociological or economic planning models which dominated subsequent urban planning in the twentieth century. Nevertheless, the concept of Civic Art had not been entirely lost by the middle of the twentieth century and formed a bridge to postmodern theories which also pay heed to the design of the city.[1]

This chapter explores the framework of Civic Art and its associated literature in Western Europe and North America. It addresses the modernist critique that Civic Art dealt only with the city's aesthetic aspects and demonstrates the limitations of this interpretation.

Stadtbaukunst in Germany and Austria

The concept of *Stadtbaukunst* refers to combining the value of beauty with various other aspects of the city. The roots of this lay in the concept of *Embellissement,* which, in the context of the French Enlightenment, was supposed to promote the economy, citizens' prosperity, hygiene, urban infrastructure and the beauty of the city. It was seen as sufficiently important for a professor of *Stadtbaukunst,* Heinrich Gentz, to be appointed at the newly founded Berliner Bauakademie as early as 1799.

Probably the most influential foundational text of modern urban design, Camillo Sitte's 1889 treatise *Der Städtebau nach seinen künstlerischen Grundsätzen* (Urban design according to its artistic principles) (Fig. 2.1) was also the impetus for modern *Stadtbaukunst* theory. Sitte juxtaposed urban design and art programmatically in the title and sought to bring an aesthetic aspect to the primarily technical engineering-focused urban development of the time. In doing so, he put the perceptual subject at the centre of his theory of urban design and called for corresponding urban spaces which were pleasant to experience. However, he did not by any means aim to reduce urban design to its artistic aspects. In the *Städtebau* treatise, he cited the technical, hygienic and economic conditions of the city as fundamental aspects which should not be ignored.[2]

Sitte's comprehensive framework becomes even clearer in the journal *Der Städtebau,* which he published with Theodor Goecke: this was the world's first specialist journal focusing on urban design.[3] In 1904, in the foreword to the first volume, Sitte and Goecke explained their comprehensive understanding of urban design:

> Urban design unifies all the technical and fine arts in a large closed-off ensemble; urban design is the monumental expression of true civic pride, the seeding ground of true love of *Heimat*; urban design deals with transport, has to provide the foundations for healthy and comfortable lives for modern people, the vast majority of whom now already live in the cities; must ensure the expedient accommodation of industry and trade and help to reconcile social extremes.[4]

Even the founding father of urban design, whose focus is on aesthetics, is therefore very far from reducing urban design as Civic Art to the aesthetic.

An eloquent testimony to a comprehensive understanding of Civic Art which focuses on design but also encompasses all other aspects of the city is provided by the lecture series *Städtebauliche Vorträge,* organized and subsequently published by Joseph Brix and Felix Genzmer in 1908–20 at the Technische Hochschule in Berlin-Charlottenburg, comprising 63 lectures covering almost as many topics.[5] In his introductory lecture, Brix described urban design as a multidisciplinary task: it is of an 'economic, hygienic, structural engineering-related [. . .], transport-related, administrative law-related, as well as legislative, and not least artistic, nature'.[6] However, only through interaction with a wide range of topics across the entire series does it become clear how important aesthetic appearance was held to be in the context of other aspects of the city. Here, artistic, design, social, economic, political, legal, hygienic, transport-related, infrastructural, climatic, geographical, historic and heritage issues are dealt with – the range of urban design aspects, and always in relation to one another.

Beginning at the same time, the *Groß-Berlin* competition of 1908–10 is significant for the problems of Civic Art in the early twentieth century and the relationship between design and other aspects. The competition sought solutions 'to promote transport as well as beauty, public health and economy'.[7] The central requirements therefore lay in transport, aesthetics, hygiene and economics – and all had to be addressed. Werner Hegemann emphasizes the close relationship of these areas in his major urban design catalogue of 1911 on the occasion of the presentation of the competition results: 'However, the overcoming of technical and economic aspects must not be separated in terms of time from artistic aspects; both have to be tackled at the same time.'[8] Only a multidisciplinary approach which also included art could solve the current problems of the major cities. In this context, the aesthetically compelling designs of

DER

STÄDTE-BAU

NACH SEINEN

KÜNSTLERISCHEN GRUNDSÄTZEN.

EIN BEITRAG ZUR LÖSUNG
MODERNSTER FRAGEN DER ARCHITEKTUR UND MONUMENTALEN
PLASTIK UNTER BESONDERER BEZIEHUNG AUF WIEN

VON

ARCHITEKT
CAMILLO SITTE
REGIERUNGSRATH UND DIRECTOR DER K. K. STAATS-GEWERBESCHULE IN WIEN.

MIT 4 HELIOGRAVUREN UND 109 ILLUSTRATIONEN UND DETAILPLÄNEN.

WIEN 1889.
VERLAG VON CARL GRAESER.
I. AKADEMIESTRASSE 26.

FIGURE 2.1 Camillo Sitte, 1889, *Der Städtebau nach seinen künstlerischen Grundsätzen,* Vienna.

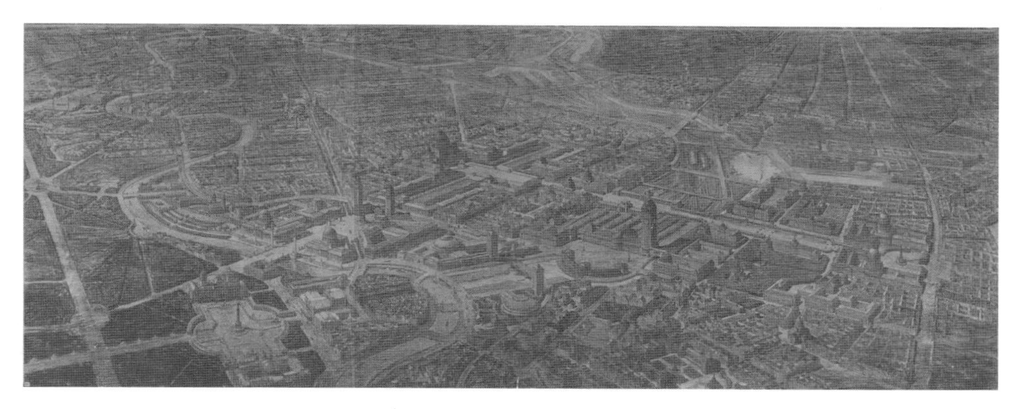

FIGURE 2.2 Bruno Schmitz, 1910, competition entry for Greater Berlin. (Architekturmuseum TU Berlin.)

Bruno Schmitz, Otto Blum, and Havestadt & Contag should be emphasized: this in no way involves purely subjective-artistic creations, as they are often described, but rather an attempt to mould the various aspects of the modern metropolis into a convincing form (Fig. 2.2). In the explanatory text, transport-related, economic and hygienic aspects thus take up by far the most space – and, in particular the composition of the competition team (an architect, an engineer and a construction firm) makes clear the multidisciplinary approach taken to the planning of the metropolis, even if artistically sophisticated forms resulted from this.

In 1920 three key publications appeared which all included *Stadtbaukunst* in their titles: the journal *Stadtbaukunst alter und neuer Zeit,* published by the art historian Cornelius Gurlitt and the architect Bruno Möhring; the handbook *Stadtbaukunst. Geschichtliche Querschnitte und neuzeitliche Ziele* by the art historian Albert Erich Brinckmann; and *Sechs Vorträge über Stadtbaukunst* by the architect Theodor Fischer.[9] In combination, these three publications paradigmatically demonstrate what *Stadtbaukunst* stood for in the early modern planning period: it was a project based on various disciplines which brought together tradition-conscious and avant-garde movements (as in the journal *Stadtbaukunst,* which contained the supplement 'Frühlicht', edited by Bruno Taut), in which historical research contributed to the solution of current problems (as in Brinckmann's *Stadtbaukunst*) and in which expert knowledge was communicated to the public (as in Fischer's *Stadtbaukunst*).

With his *Stadtbaukunst,* Brinckmann presented the first comprehensive history of urban design. While he did not overlook historical solutions to the practical problems of his time, he also emphasized the formal–aesthetic side of urban development: 'The programme of artistic urban design is to establish a relationship between building dimensions and spaces and develop them together.'[10] As a pupil of the art historian Heinrich Wölfflin, Brinckmann did not, however, advocate a strictly formalistic art historical approach since in the first sentence of his *Stadtbaukunst,* he describes the multidisciplinary nature of his subject: he mentions 'the social and economic conditions of urban development', which he now wants to complement with a 'history of the form of the cities'.[11] His overall presentation of the history of urban development is characterized by the view that the city does not develop its form autonomously; rather, it is influenced to do so by different factors.

The architect Fischer formulated his subject in precisely the same manner in his *Sechs Vorträge über Stadtbaukunst.* In the tradition of Gottfried Semper, he referred to his interest as 'practical aesthetics [. . .], broadly based on experience'.[12] The aesthetic issue was thus linked back to the practical tasks with which urban design had to deal. He comprehensively described the subject of *Stadtbaukunst* as 'the vast field of urban design' and continued:

> It is amazing what one has to deal with when one considers the economic and technical principles of the subject! The complexity of land policy with the barbed wire of the Federal Civil Code and the wolf's lair of the mortgage sector; then the steel walls of the building inspection department and communication trenches of public bureaucracy. On top of this, the constraints and fiscal impossibilities of transport policy and the surprises of population movement. The economist, the statistician, the bank and insurance expert, the administration, building inspection department and traffic civil servants, they all want to and have to talk to one another before the engineer has his say. After all, his own field is already wide enough, much too large for *one* expert. The division of labour goes further: first of all, the hygiene expert sets out fundamental conditions, the fulfilment of which is usually placed in the hands of civil engineers; it is their job to supply the localities with water and take away effluent. The engineer also provides the city with light and power, the engineer builds roads and provides routes and rails for wheeled vehicles. [. . .] However, someone has been forgotten, the person who really matters if the end result, that which will remain, the city, is to be something worthwhile: the architect building the houses – it is for their sake, the homes of the people, that everything else has been supplied and undertaken. His role is to provide advice and be involved in the process and, furthermore, his role is to conduct the process because it is his activity which is the intention and final objective of the whole undertaking.[13]

According to Fischer, *Stadtbaukunst* emerged from the interplay of economic, political, legal, social, technical and cultural questions, all represented by a variety of professions. These aspects are, in theory, brought together in the configuration of the city and its houses, a crowning role which falls to the architect.

In 1929 the architect and art historian Paul Zucker presented a history of the form of the city with his publication *Entwicklung des Stadtbildes. Die Stadt als Form*. In his introduction, he commented that 'the prerequisites of urbanisation have already been sufficiently analysed from a social, economic, housing policy and transport perspective'. He now wanted to add to this the 'discussion and analysis of the formal appearance of the embodied city'.[14] However, his history of the form of the city in no way ruled out other factors. In contrast, he expressly stated that an appropriate understanding of the city can be arrived at only by considering various factors together:

> Any consideration of the organism of the city must always proceed from a dual perspective: on the one hand from the aesthetic view of the city as a unique work of art which stands in space with its own internal rules, as well as from the biological perspective which regards the city as a living, continuously developing organism which is subject to social, hygienic, economic and technical rules, and as such stands in time.[15]

Zucker was to maintain this view throughout his life. It also characterized the symposium 'New Architecture and City Planning' that he organized in New York in 1944, at which Sigfried Giedion emphasized 'The Need for Monumentality', but Carol Aronovici also talked about 'Civic Art', which is overlooked in contemporary historical accounts of this symposium. Zucker's *Town and Square: From the Agora to the Village Green,* probably his best-known work, was characterized by the idea that civic forms can retain their validity across historic changes[16] – making him at this time one of the rare voices against the choir of those calling for innovation in the name of the contemporary.

L'art urbaine in France and Belgium

In France and Belgium, the terms *l'art urbaine* or *l'art de bâtir des villes* were used when referring to giving an artistic dimension to technical engineering-related urban planning in the twentieth century. One pioneering publication in this regard was the 1893 *Esthétique des villes* by Charles Buls, the mayor of Brussels.[17] In this slim 40-page pamphlet he countered the view that he wanted to place aesthetics above all other considerations in urban development. He used individual chapters to deal with the *point de vue technique,* the *point de vue esthéthique* and the *point de vue archéologique.* The historic–aesthetic aspect also played a fundamental role in the context of all practical requirements as, in the appearance of the city, history is preserved and communicated to current and future generations – this is the real political interest of the politician Buls.

In his 1901 work *L'Esthétique de la rue,* the writer Gustave Kahn developed an almost encyclopaedic aesthetic of the street.[18] His observations attempted to record the appearance of the street in all its historical, functional and formal scope in order to then promote a uniform street aesthetic, which had to be shaped in a conscious fashion.

Another writer who dedicated himself to the beauty of the city was Emil Magne. The title of his work, published in 1908, itself underlined the variety of issues he intended to deal with: *L'Esthétique des villes. Le décor de la rue, les cortèges, marchés, bazars, foires, les cimetières, esthétique de l'eau, esthétique du feu, l'architectonique de la cité future* (The aesthetics of cities. The decoration of streets, processions, markets, bazaars, fairs, cemeteries, aesthetic of water, aesthetic of fireworks, architecture of the future city). It was, above all, the colourful activity of city life in its many aspects from which the appearance of the city was composed. He justified his assessment of existing city spaces by referring to Sitte's paradigm and, like that author, demanded that urban architecture be more focused on its context.[19] His extremely colourful and graphic descriptions of city life also convincingly directed the reader towards his call for a more diverse orchestration of the city.

In 1913 Robert de Souza delivered the first comprehensive work from an expert perspective with the rather peculiar-sounding title *Nice capitale d'hiver. L'Avenir de nos villes. Études pratiques d'esthétique urbaine* (Nice, winter capital. The future of our cities. Practical exercises in urban aesthetics). The aesthetics of the city were the focus of his interest but were viewed in close relationship with other factors:

> It becomes clear that economic, hygienic, administrative, financial, commercial and social questions are the foundation of an effective aesthetic. To be constructive, the aesthetic must approach each of the problems which these questions raise; it must not neglect one of them.[20]

One of the main objectives of all these efforts was a beautiful cityscape. In this regard, de Souza quoted Stübben, with emphatic capitalization: '"The beauty of the cities' appearance is an essential condition of their existence." Their EXISTENCE! . . . really, their EXISTENCE . . .'[21]

In 1915 *Comment reconstruire nos cités détruites* by Donat-Alfred Agache, Marcel Auburtin and Edouard Redont appeared as a handbook in the context of the First World War. They cited '*l'hygiène et l'esthétique*' as the main responsibilities of urban design.[22] Beauty did not mean lack of comfort and uncleanliness; it was a result of 'harmonious proportions, picturesque silhouettes, suitable use of materials, etc.'[23] To underline their attitude that beauty was not a luxury for the rich in better times, they quoted the highly esteemed Parisian city architect Louis Bonnier: 'The aesthetic is not a luxury for the people, but a right and a need in the same respect as hygiene.'[24] Beauty and hygiene as equal requirements, as well as beauty as a right for workers, typified the attitude of early modern French urbanism to urban design.

The handbook by the Belgian landscape architect Louis van der Swaelmen (*Art Civique,* 1915) translated the English term 'Civic Art' into French.[25] George Burdett Ford's *L'urbanisme en pratique. Précis de l'urbanisme dans tout son extension* (Urbanism in practice. Compendium of urbanism in its entire extensions) of 1920 was the first to feature the term 'urbanism' in the title, and this was followed in 1923 by Edmond Joyant's *Traité d'urbanisme* (Treatise of urbanism). All these handbooks also contained a separate chapter on the beauty of the city.

The close link between the history of urban design and urban planning was embodied in France by Marcel Poëte, who, as a historian and librarian, published a history of the urban development of Paris and a general history of urban development but was also involved in the expansion plan for Paris of 1913.[26] In 1929 he presented a comprehensive discussion of urban design in his *Introduction à l'urbanisme* (Introduction to urbanism). With this book he planned a systematic portrayal and general history of urban design, which nevertheless did not go beyond antiquity. Urban design was simultaneously art and science and covered 'various disciplines: economy, geography, history and others'.[27] These different disciplines corresponded to the different aspects of urban design, 'to historic facts you have to add geographic, geological and economic facts'. All the aspects were also closely linked to one another: 'the economic characteristics help to explain the social characteristics, in the same way as the political and administrative characteristics are linked to them'.[28]

Arte di costruire le città in Italy

The discussion of urban design in Italy did not initially develop its own accent. The engineer and archaeologist Ugo Monneret de Villard translated the German term *Stadtbaukunst* into Italian for the work *Note sull'arte di costruire le città* (Notes on the art of building cities), which he published in 1907, and advocated the ideas of Charles Buls and Camillo Sitte.[29] In two articles published in 1913, Gustavo Giovannoni further developed Buls's ideas of maintaining the historic surroundings of urban monumental structures into his own theory on how to deal with old towns, although it was not until 1931 that he presented a general work on urban design.[30] In both articles, which appeared in 1913 in quick succession in the same journal, Giovannoni discussed the issue of dealing with historic cities. Radical modernization ran counter to the appreciation of existing cityscapes, an appreciation which had aesthetic, as well as historic and political, motivation. In Giovannoni's view, the issue of the overall image played a central role, for which he coined the term *ambiente:* not just the individual monumental structure but also the entire cityscape had a value as an artistic structure, as a historic witness and as an expression of belonging.

The first Italian handbook to openly recognize its sources in its title, *Costruzione, trasformazione e ampliamento delle città. Compilato sulla traccia del Der Städtebau di J. Stübben* (Construction, transformation and extension of cities. Compiled from Der Stadtebau by J. Stubben) by Aristide Caccia in 1915, was nothing more than a pragmatic compilation of Stübben's work, which had appeared a quarter of a century before. The combination of beauty and functionality characterized Marcello Piacentini's first publication on the development of modern Rome of 1916, a concept which was to be reflected in all his subsequent urban design work.[31] Piacentini, with his urban constructions and his urban design plans, shaped Italian Civic Art up to the middle of the twentieth century. His designation of an *edilizia cittadina* in 1922[32] can be understood as the counterpart of the German term *Stadtbaukunst* or the English 'Civic Art'. Central to this concept is the idea that the tasks of modern urban development will lead to the renewal of architecture. In a variety of manners, Piacentini therefore emphasized the subordination of the individual to the collective: he advocated, on the one hand, the subordination of the individual designer to the style of the time – an *impersonalitá* of the architect – and, on the other hand, the assignment of all constructions into the context – an *ambientismo* of architecture which went back to the deliberations of Giovannoni. Just like all the other authors of Civic Art, he exposed the complexity of building, which, as an art, had to encompass and combine aspects of aesthetics, morals, sociology, hygiene and security.

Civic Art in Great Britain

In the English-speaking world, it was the term 'Civic Art' which combined the civic–societal with the aesthetic–artistic. In the UK, the discourse on a form of urban development supported by citizens who aimed to create a beautiful city arose from the Arts and Crafts movement. For example, in 1896 William Richard Lethaby gave a lecture to the Arts and Craft Exhibition Society on 'Beautiful Cities' in which he vehemently opposed a purely technical and economic approach to urban design.[33]

It was precisely these notions to which the main protagonist of English urban development, Raymond Unwin, referred in the first chapter, 'Of Civic Art as the Expression of Civic Life', in his 1909 *Town Planning in Practice:* 'Remembering then that art is expression and that civic art must be the expression of the life of the community, we cannot well have a more safe practical guide than Mr. Lethaby's saying that "Art is the well-doing of what needs doing".'[34] Unwin asserted that beauty is one of the fundamental tasks of urban design and is supposed to give the community a voice. In 1911 Unwin once again emphasized the independent role of beauty in urban design in an article relating to the debate on 'Town Planning: Formal or Irregular'. He argued that 'in town planning, at any rate, beauty is intimately associated with use, with fitness for purpose and function, but it is not the same thing. It is not enough to satisfy the use and trust that by chance beauty will result.'[35] According to Unwin, beauty is associated with function but is in no way identical to it. It is not an automatic product of the fulfilment of purpose; rather, it requires its own formal considerations. It is significant for this period that he propagated this under the term 'Town Planning', which at this time still encompassed the design-related aspects of urban planning.

One important handbook was Thomas Mawson's *Civic Art,* which appeared in 1911 (Fig. 2.3). Here, the author, who primarily worked as a landscape architect, advocated a public Civic Art for a 'communal or civic life, of which every man may be a member'.[36] However, he in no way saw 'Civic Design' as purely aesthetic: 'It covers such a vast field of legal, economic and social

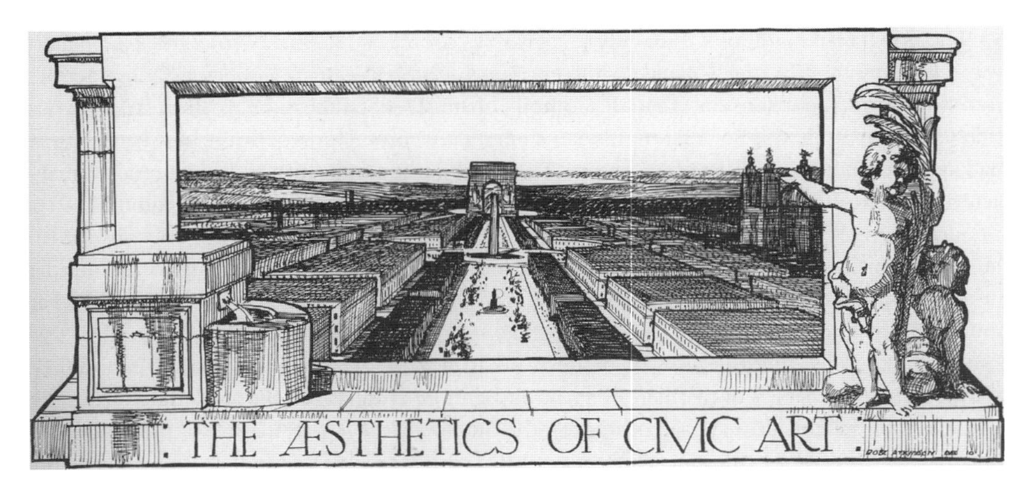

FIGURE 2.3 Thomas Hayton Mawson, 1911, *Civic Art: Studies in Town Planning, Parks, Boulevards, and Open Spaces,* London.

subjects, including the scientific or technical departments of civil engineering, architecture, forestry and horticulture.'[37] Nevertheless, the beauty of the city demanded separate consideration:

> Although civic beauty so often rests on good planning it is not maintained that a well-planned city is necessarily a beautiful city in the ordinarily accepted sense of the word; the strict meaning of beauty always carries with it the expressive sense of fitness of the visible exterior to fulfil its function.[38]

Mawson's urban aesthetics clearly differed, however, from the more picturesque views of the English Garden City movement: Mawson propagated the ideal of large-scale geometric design in the style of Baroque gardens and cities, as contemporarily also advocated by the French École des Beaux-Arts and the US City Beautiful movement.

Alongside a large number of other authors, particular emphasis should be given to the English architect Arthur Trystan Edwards, who studied and presented in detail an appropriate aesthetic approach to urban design. The hierarchy of public and private buildings as well as the harmonic and cityscape-forming conjunction of private buildings are the themes of his essays collected in *Good and Bad Manners in Architecture* from 1924.[39] In striking practical–aesthetic studies, Trystan Edwards addressed the problem of Civic Art. He did not want to simply provide a handbook-type overview but to deliver specific lessons for the design of urban buildings. In a direct translation of the term Civic Art from a societal–political to an architectonic–aesthetic understanding, he views the appearance of civic building from the societal paradigm of manners. His criticism was of 'selfish buildings', which paid no heed to the cityscape but were, above all, attention-seeking and focused on the profits of their builders and the reputations of their designers. Using sketches, Trystan Edwards drew up and tested rules by which good urban architecture could take account of both the hierarchy between public and private buildings and the appropriate relation between private buildings in a street (Figs 2.4, 2.5). The general principle is harmonious interaction in contrast to solipsistic cacophony.

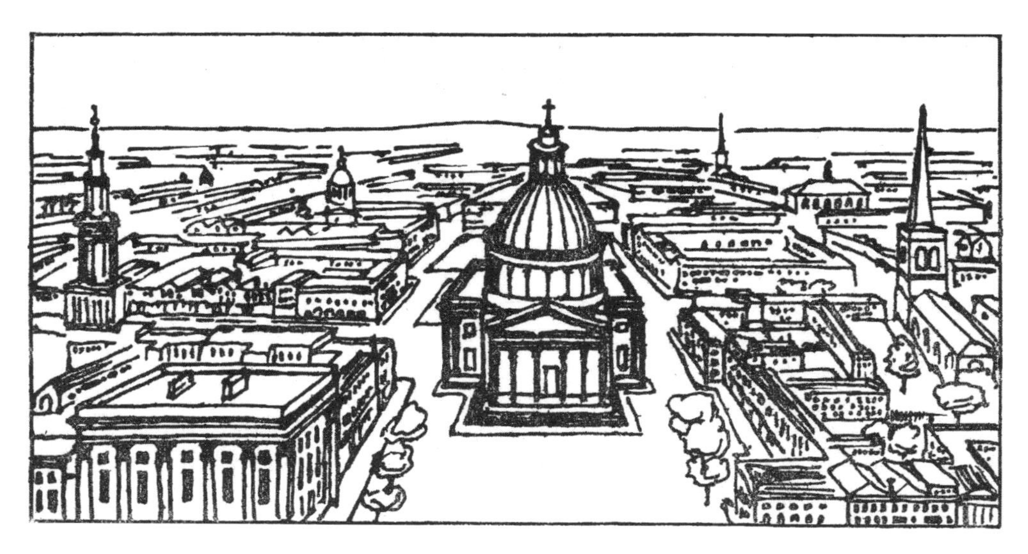

FIGURE 2.4 Arthur Trystan Edwards, 1924, *Good and Bad Manners in Architecture,* London. 'This first drawing depicts an imaginary city of the old-fashioned type, in which the principal public buildings are given a forma pre-eminence' (p. 2).

FIGURE 2.5 Arthur Trystan Edwards, 1924, *Good and Bad Manners in Architecture,* London. 'This drawing shows the same city after certain modern influences have been operating for a number of years. Several very selfish commercial buildings have now arrived, with the result that the dome of the cathedral now no longer holds undisputed sway' (p. 5).

Even more originally, Trystan Edwards implemented his practical urban aesthetics in a series of articles under the title 'What the Building Said', which appeared in 1926 in *The Architectural Review;* in these articles the buildings on a street discussed their mutual advantages and disadvantages in a fictional conversation. Far removed from any dogmatism, rules of Civic Art were developed on the basis of common sense in a critical argument of statement and counterstatement. Here, societal behaviour is superimposed directly onto the aesthetic behaviour of urban houses, with the term 'Civic Art' being implemented in a literal manner. Trystan Edwards's emphasis on aesthetic issues was directly influenced by Geoffrey Scott's seminal book *The Architecture of Humanism,* in which he had developed the inherent formal value of architecture against the effects of non-architectonic conditions.[40]

In 1925, with his book *The Art of Town Planning,* the English architect Henry Vaughan Lanchester once again underlined the comprehensive understanding of early British town planning in the sense of Civic Art which had also shaped Unwin's fundamental handbook *Town Planning in Practice* of 1909. In addition to all the practical requirements, and taking into account all the specific aspects, the beauty of the city was also a central responsibility of urban planning; only when this was the case would town planning become a 'fine art'.[41]

Thomas Sharp provides appropriate evidence of the long-lasting effect of this both aesthetical and functional urban design as Civic Art in the UK. In his widely distributed 1940 handbook, *Town Planning* (Fig. 2.6), he defined a 'good town' as 'a utility for collective living'.[42] This functional specification of the city does not give the whole picture but is also complemented with an aesthetic aspect:

> It is necessary that his town should be more than merely comfortable to live in. It must be pleasing to look at. It cannot, in fact, in the long run be comfortable to live in unless it is also visually satisfying. Beauty, or at the least order and seemliness, is as necessary to a civilized life as health and convenience and mere organisational functionalism.[43]

Sharp's texts and urban plans, which argued for townscape aesthetics and put these into practice, retained their influence into the post-Second World War period and formed a direct bridge to the Townscape movement; the term 'townscape' can be traced back to Sharp, although it is more usually associated with Gordon Cullen.

Civic Art in the US

In the period under discussion here, America developed into an influential centre of urban planning and Civic Art. Against the background of the City Beautiful movement, which had been triggered by the uniform development of the White City at the World Exposition in Chicago in 1893, urban improvement and urban beautification became a responsibility to be borne by the citizens, which resulted in plans for all of the more important American cities. One of the first authors was Charles Mulford Robinson, who drew up 14 plans for American cities. In the foreword of his main work, *The Improvement of Towns and Cities,* he described the purpose of this 1901 book as a 'battle for urban beauty'.[44] While the beauty of the city had once been a self-serving matter of the aristocracy, it now represented a value of democratic society. Robinson's work *Modern Civic Art* from 1903 was a popular variant of his textbook. He thus lived up to the claim of general comprehensibility and distribution, which also corresponded with his ideal of urban planning. Under the title 'What Civic Art Is', he stated, 'It is municipal first of all'.[45] Civic Art is not just for art and the artist; it is also for the city and the citizen: 'Civic art is essentially public art'.[46]

The comprehensive understanding of the topic within the City Beautiful movement was reflected in contemporary publications. The impressive list of American urban design

TOWN AND COUNTRY

1. Town : Elgin.
2. English country pattern.

FIGURE 2.6 Thomas Sharp, townscape and landscape, 1940, *Town Planning,* London.

literature ranges from the key planning reports for Washington and Chicago, which were tailored to specific planning but contained broad historical and general sections,[47] via the religious publications of Josiah Strong,[48] the sociological publications of Charles Zueblin,[49] the political publications of Delos F. Wilcox, Frederic Clemson Howe, and Benjamin Clarke Marsh and George Burdett Ford,[50] and the technical engineering publications of Frank Koester,[51] through to the planning literature of John Nolen and Nelson P. Lewis.[52] In 1922 *The American Vitruvius* by Werner Hegemann and Elbert Peets (the full title of which continued *An Architects' Handbook of Civic Art*) finally appeared as a codifying handbook. In this publication, model urban development solutions from 2,500 years of urban development history were presented by a German economist and an American landscape architect. However, the intention was not to present a historical account, but rather a useful collection of good solutions, as the authors state in the foreword: 'The purpose was not to make a history of civic art, [but] the compilation of a thesaurus, a representative collection of creations in civic art.'[53]

Looking back, Charles Moore identified the comprehensive urban objectives of the City Beautiful in an article 'The City as a Work of Art' in 1931: 'What is meant by Delight in planning a city? Nothing less than that the entire city shall be planned as a Work of Art.'[54] The 'New Architecture and City Planning' symposium (1944) organized and edited by Paul Zucker, at which Giedion and Louis Kahn presented their thoughts on 'A New Monumentality', was evidence of the continued tradition of Civic Art in America: here, lectures entitled 'Civic Art' and 'The City as a Work of Art' were also given.[55] A further cornerstone of this history is the Civic Art Conference which Christopher Tunnard organized in 1952 at Yale University. Speakers such as Henry Hope Reed and Hans Blumenfeld emphasized the role of the ornamental and human scale in urban development – topics which ran counter to the prevailing functionalistic urban planning. One of those in the audience was Kevin Lynch, whose influential work *The Image of the City* was to appear a few years later.[56]

Civic Art in practice

Such a theory of Civic Art was open to all sides and was not closed off from practice. It is characteristic of Civic Art theory of this time that it was substantially based on practical experience and was carried out by practitioners. As a result of this close association, the urban development practice of this time was also able to serve as a litmus test for the suitability of Civic Art theory. In fact, the results of this test were very positive since the advocates of theoretical Civic Art were those practical urban designers who created beautiful, high-quality, functional and long-lasting urban areas, which today still represent the most popular – and therefore most expensive – areas in their cities because of the quality of their urban design.

In Germany, this includes, for example, Schwabing in Munich, which originated from the plan by Fischer and, in addition to stately urban residential houses, contains pioneering public buildings of the master himself. Charlottenburg in Berlin is included, where at this time the Hobrecht plan was filled with the homely yet metropolitan residential buildings of Albert Gessner – still the most popular city apartments because of their generous floor plans and room dimensions, their solidity and their external beauty. Also included are the urban areas by Fritz Schumacher in Hamburg, where he also succeeded in bringing an urban and space-configuring, as well as long-lasting and beautiful, design to social residential buildings.

Internationally, these were also the years of successful Civic Art practice. It was the time when Eliel Saarinen drew up his comprehensively configured metropolitan visions for Helsinki, which resulted in urban quarters such as Töölö. It was the time when Hendrik Petrus Berlage conceived the urban expansion of Amsterdam as artistic metropolitan events such as the coherent and diverse city quarters in Amsterdam-South or Amsterdam-West. In Paris, the Boulevard Periphérique on the former fortifications was developed with metropolitan blocks and street spaces – the last expansion of Paris which was conceived in continuity with the existing city. In Vienna, residential buildings with a metropolitan design, the courtyards (*Höfe*), were constructed both in the city and on its edges. Even London, the city of the Garden Suburbs, had its own inner-city residential blocks which with 'modern Georgian' further developed the *genius loci* of the metropolis, both in the construction of social housing and on the free market.[57] In Italy too, the classical urban development tradition was continued in modern fashion with the Milanese residential buildings of Novecento or the square constructions of Marcello Piacentini in Bergamo, Brescia, Bozen and Genoa. The Civic Art of these years was able to bring about a practice which was so successful in the long term because it had at its disposal an appropriately complex theory.

However, selected examples of the reconstruction after the Second World War are also shaped by the concept of Civic Art. In Germany, these include, for example, the traditionalist plans of Paul Schmitthenner, Karl Gruber, Adolf Abel or Emil Steffann for Lübeck, Mainz, Darmstadt, Wiesbaden or Siegen, which, however, in almost all cases were not implemented. The plans by Karl Meitinger for Munich, in which 'our beloved Munich is recreated in a new guise, but in the old spirit', enjoyed more success.[58] In the German Democratic Republic, the constructions of Socialist Realism, the most famous of which is Stalinallee in Berlin, can be traced back directly to the tradition of Civic Art (compare Fig. 2.7 with Fig. 2.8). In France, Auguste Perret in Le Havre and Fernand Pouillon in Marseilles developed civic spaces which work with the traditional elements of Civic Art. In the UK, the London Plans of the Royal Academy Planning Committee, chaired by Edwin Lutyens, lay in the tradition of the large-scale urban plans of early Civic Art. The plans of Thomas Sharp for Exeter, Durham, Oxford and Salisbury are highlights of a style of planning which focuses on the shape of the city. Examples such as the construction of Praça do Areeiro in Lisbon by Cristino da Silva also bear witness to the survival of Civic Art in the times of functionalism.

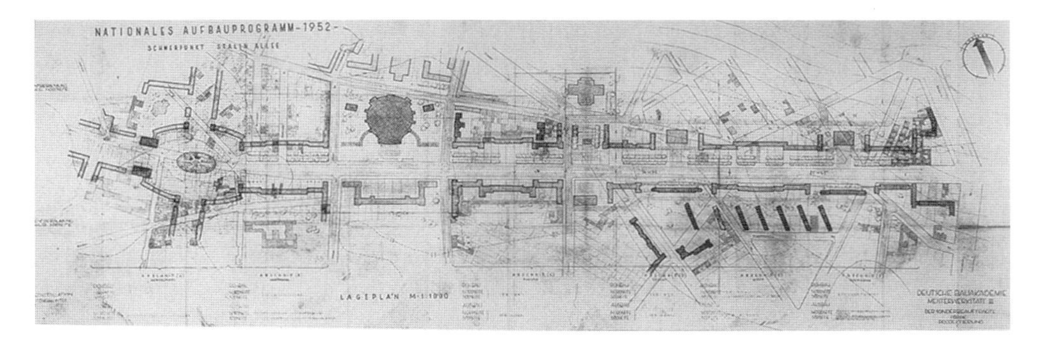

FIGURE 2.7 Egon Hartmann, Richard Paulick and others, 1951, Stalinallee, Berlin.

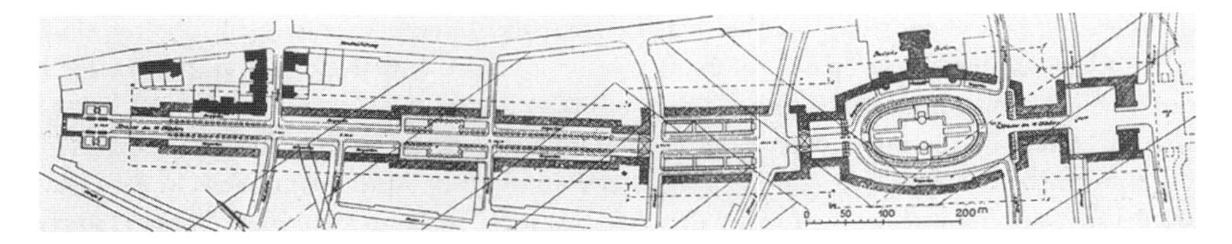

FIGURE 2.8 Straße des 18. Oktober in Leipzig, 1916, in Werner Hegemann and Elbert Peets, 1922, *The American Vitruvius,* New York.

Conclusion

If transport, hygienic housing and recreational spaces represented the most urgent practical problems of urban development in the early twentieth century, aesthetics were, however, never neglected: lack of beauty was a trigger for comprehensive planning, and a beautiful cityscape was regarded as the highest objective of urban design. The development of modern urban design theory can also be described as a multidisciplinary discursive process in which a large number of disciplines were involved in founding a science of urban design. These included engineers such as Ildefonso Cerdà, architects such as Hendrik Petrus Berlage and Daniel Hudson Burnham, economists such as Rudolf Eberstadt and Werner Hegemann, sociologists such as Max Weber and Robert Park, philosophers such as Georg Simmel, political scientists such as Frederic Howe, natural scientists such as Patrick Geddes, historians such as Marcel Poëte, art historians such as Albert Erich Brinckmann and art critics such as Karl Scheffler – to mention only the most important.[59]

There was also no strict division into specific fields: these authors usually had a comprehensive understanding of the city which sought to solve design-related problems related to economic, political, social, technical, legal, sanitary and other problems. As early as 1922, however, Hegemann and Peets criticized the drifting apart of individual areas and an increasingly solely social and technical understanding of urban planning which neglected design aspects when they wrote in *American Vitruvius:* 'Indeed, the authors feel that the young profession of city planning is drifting too strongly in the directions of engineering and applied sociology.'[60] It is noteworthy that this criticism was voiced long before urban planning was to be restricted almost exclusively to engineering and sociology, but also, importantly, that it did not originate with an art historian or architect but with an economist.

In fact, the comprehensive concept of Civic Art was sidelined by advocates of various avant-garde movements who were happy to reduce the city to a few factors, and if possible to only one factor; this helped their theories to take hold and, for example, led to the simplifying reduction of the city to four functions in the Athens Charter of 1933 and 1943. This disproportionate simplification in the name of innovation was, however, not an essential quality of the *Zeitgeist:* even at the time it was picked up and heavily criticized by observant contemporaries. One example of this is the sharp-tongued discussion of Le Corbusier's *The City of Tomorrow* of 1929 by the English architect and architectural theoretician A. Trystan Edwards, discussed earlier. His criticism focused on the reductionist urban view of Le Corbusier:

> It is much too easy to design ideal or Utopian cities if the artist concentrates upon two or three of the factors to be considered and rejects all others . . . M. Le Corbusier's solution

> is to do away with the complexity. This complexity, however, is part of the subject of civic design. The modern great city is like a large orchestra which often plays an inferior piece of music, and in which the instruments themselves may occasionally even be out of tune. It is the business of a reformer to improve the music and the instruments, but not to cut the range of the orchestra, nor the number of musical effects that are aimed at by it. M. Le Corbusier has not the patience to attempt this, but substitutes for this orchestra a single tin whistle with about five notes, with which he plays a perfectly rhythmical tune. But it is not enough.[61]

Just as one cannot play an Elgar symphony with a tin whistle, the reduced range of instruments of the avant-garde is incapable of formulating an adequate understanding of the complex phenomenon of the city.

Despite this, the whistle players at first had their noses in front as their instruments were not only louder but also much easier to play and understand. The role of design in urban development was therefore not played down only by engineers, sociologists, economists and political scientists – rather, in most cases, it was avant-garde architects themselves who sought renown in external factors and other disciplines and denied the independent role of design, form and aesthetics in city and architecture under the dictum of functionalism. The concept of Civic Art was thus temporarily sidelined – and, with it, cityscapes with a sophisticated and beautiful design.

Notes

1. Dethier, J., Guiheux, A. (eds) (1994) *La Ville. Art et architecture en Europe 1870–1993,* Paris: Editions du Centre Pompidou; Sonne, W. (2003) *Representing the State: Capital City Planning in the Early Twentieth Century,* Munich, London, New York: Prestel; Sonne, W. (2003) '"The entire city shall be planned as a Work of Art." Städtebau als Kunst im frühen modernen Urbanismus 1890–1920', in: *Zeitschrift für Kunstgeschichte,* 66(2): 207–236; Sonne, W. (2010) 'Stadtbaukunst alter und neuer Zeit', in: Mäckler, C. and Sonne, W. (eds), *Dortmunder Vorträge zur Stadtbaukunst,* vol. 2, Sulgen: Niggli, pp. 10–29.
2. Sitte, C. (1889) *Der Städtebau nach seinen künstlerischen Grundsätzen,* Vienna: Graeser, preface.
3. Its subtitle was *Monatsschrift für die künstlerische Ausgestaltung der Städte nach ihren wirtschaftlichen, gesundheitlichen und sozialen Grundsätzen* (Monthly periodical for the artistic design of cities according to economic, hygienic and social principles).
4. All translations are the author's own.
5. Brix, J., Genzmer, F. (eds) (1908–20) *Städtebauliche Vorträge aus dem Seminar für Städtebau an der Königlichen Technischen Hochschule zu Berlin,* Berlin: Wilhelm Ernst & Son.
6. Brix, J. (1908) 'Aufgaben und Ziele des Städtebaus', in: Brix and Genzmer, *Städtebauliche Vorträge,* vol. 1, no. 1, p. 11.
7. 'Wettbewerb um einen Grundplan für die Bebauung von Gross-Berlin', *Der Baumeister,* 7(2), 1908, p. B18; also in *Wochenschrift des Architekten-Vereins zu Berlin,* 3(50), 1908, p. 275.
8. Hegemann, W. (1911) *Der Städtebau nach den Ergebnissen der allgemeinen Städtebau-Ausstellung in Berlin nebst einem Anhang: Die internationale Städtebau-Ausstellung in Düsseldorf,* vol. 1, Berlin: E. Wasmuth, p. 129.
9. Gurlitt, C., Möhring, B. (eds) (1920–28) *Stadtbaukunst alter und neuer Zeit,* 8 vols.; Brinckmann, A. E. (1920) *Stadtbaukunst. Geschichtliche Querschnitte und neuzeitliche Ziele,* Berlin: Akademische Verlagsgesellschaft Athenaion; Fischer, T. (1920) *Sechs Vorträge über Stadtbaukunst,* Munich, Berlin: R. Oldenbourg.
10. Brinckmann, *Stadtbaukunst,* p. 134.
11. 'Formgeschichte der Städte', foreword in ibid.
12. Fischer, *Sechs Vorträge,* p. 5.

13. Ibid., pp. 6–7.
14. Zucker, P. (1929) *Entwicklung des Stadtbildes. Die Stadt als Form,* Munich, Berlin: Drei Masken, p. 7.
15. Ibid., p. 11.
16. Zucker, P. (1959) *Town and Square: From the Agora to the Village Green,* New York: Columbia University Press.
17. Buls, C. (1893) *Esthétique des villes,* Brussels: Bruyland Christophe.
18. Kahn, G. (1901) *L'Esthétique de la rue,* Paris: Charpentier.
19. Magne, E. (1908) *L'Esthétique des villes. Le décor de la rue, les cortèges, marchés, bazars, foires, les cimetières, esthétique de l'eau, esthétique du feu, l'architectonique de la cité future,* Paris: Mercure de France, p. 18.
20. De Souza, R. (1913) *Nice capitale d'hiver. L'Avenir de nos villes. Études pratiques d'esthétique urbaine,* Paris: Berger Levrault, foreword.
21. Ibid., p. 370.
22. Agache, D.-A., Auburtin, M. and Redont, E. (1915) *Comment reconstruire nos cités détruites. Notions d'urbanisme s'appliquant aux villes, bourgs et villages,* Paris: Armand Colin, p. xi.
23. Ibid., p. 7.
24. Ibid., p. 52.
25. Van der Swaelmen, L. (1915) *Préliminaires d'art civique mis en relation avec le 'cas clinique' de la Belgique,* Leiden: A. W. Sijthoff.
26. Préfecture du Département de la Seine, Commission d'Extension de Paris (1913) *Aperçu historique,* Paris: Imprimerie Chaix; Poëte, M. (1921–31) *Une vie de cité. Paris de sa naissance à nos jours,* 4 vols., Paris: Picard.
27. Poëte, M. (1929) *Introduction à l'urbanisme. L'évolution des villes,* Paris: Boivin, p. 1.
28. Ibid., p. 3.
29. Monneret de Villard, U. (ed.) (1907) *Note sull'arte di costruire le città,* Milan: Societa Editrice Technico Scientifica.
30. Giovannoni, G. (1913) 'Vecchie città ed edilizia nuova. Il quartiere del Rinascimento in Roma', *Nuova Antologia di lettere, scienze ed arti,* 48(995): 449–472; Giovannoni, G. (1913) 'Il "diradamento edilizio" dei vecchi centri. Il quartiere della Rinascenza in Roma', *Nuova Antologia,* 48(997): 53–76; Giovannoni, G. (1931) *Vecchie città ed edilizia nuova,* Turin: Utet.
31. Piacentini, M. (1916) *Sulla conservazione della bellezza di Roma e sullo sviluppo della città moderna,* Rome: Associazione Artistica Cultori di Architettura.
32. Piacentini, M. (1922) 'Nuovi orizzonti nell'edilizia cittadina', *Nuova antologia di lettere, scienze ed arti,* 57(119): 60–72.
33. Arts and Craft Exhibition Society (ed.) (1897) *Art and Life, and the Building and Decoration of Cities,* London: Rivington, Percival, pp. 45–110.
34. Unwin, R. (1909) *Town Planning in Practice,* London: Ernest Benn, p. 10.
35. Unwin, R. (1911) 'Town Planning: Formal or Irregular', *The Builder,* 101, p. 441.
36. Mawson, T. H. (1911) *Civic Art: Studies in Town Planning, Parks, Boulevards, and Open Spaces,* London: B. T. Batsford, p. 6.
37. Ibid., p. 9.
38. Ibid., p. 32.
39. Trystan Edwards, A. (1924) *Good and Bad Manners in Architecture,* London: P. Allan.
40. Scott, G. (1914) *The Architecture of Humanism: A Study in the History of Taste,* London: Constable.
41. Lanchester, H. V. (1925) *The Art of Town Planning,* London: Chapman and Hall.
42. Sharp, T. (1940) *Town Planning,* Harmondsworth: Penguin, p. 58.
43. Ibid., p. 59; Pendlebury, J. (ed.) (2009) 'Thomas Sharp and the Modern Townscape', special issue, *Planning Perspectives,* 24(1).
44. Robinson, C. M. (1901) *The Improvement of Towns and Cities, or, The Practical Basis of Civic Aesthetics,* New York: G. P. Putnam, p. viii.
45. Robinson, C. M. (1903) *Modern Civic Art, or, The City Made Beautiful,* New York: G. P. Putnam, p. 26.
46. Ibid., p. 36.
47. Moore, C. (ed.) (1902) *The Improvement of the Park System of the District of Columbia. I. – Report of the Senate Committee on the District of Columbia. II. – Report of the Park Commission,* 57th Congress, 1st Session, Senate Report No. 166, Washington; Burnham, D. H., Bennett, E. H. (1909) *Plan of Chicago,* Chicago: Commercial Club of Chicago.
48. Strong, J. (1898) *The Twentieth Century City,* New York: Baker and Taylor; Strong, J. (1907) *The Challenge of the City,* New York: Baker and Taylor.

49. Zueblin, C. (1903) *American Municipal Progress: Chapters in Municipal Sociology,* New York, London: Macmillan; Zueblin, C. (1905) *A Decade of Civic Development,* Chicago: University of Chicago Press.
50. Wilcox, D. F. (1904) *The American City: A Problem in Democracy,* New York, London: Macmillan; Howe, F. C. (1905) *The City: The Hope of Democracy,* New York: C. Scribner; Marsh, B. C., Ford, G. B. (1909) *An Introduction to City Planning: Democracy's Challenge to the American City,* New York: Marsh; Howe, F. C. (1915) *The Modern City and Its Problems,* New York, Chicago, Boston: Scribner.
51. Koester, F. (1914) *Modern City Planning and Maintenance,* New York: McBride, Nast.
52. Nolen, J. (ed.) (1916) *City Planning: A Series of Papers Presenting the Essential Elements of a City Plan,* New York, London: D. Appleton; Lewis, N. P. (1916) *The Planning of the Modern City: A Review of the Principles Governing City Planning,* New York: John Wiley.
53. Hegemann, W., Peets, E. (1922) *The American Vitruvius: An Architect's Handbook of Civic Art,* New York: Architectural Book Publishing, foreword.
54. Moore, C. (1931) 'The City as a Work of Art', *City Planning,* 7(2), p. 69.
55. Boas, G. (1944) 'The City as a Work of Art', Zucker, P. (ed.), *New Architecture and City Planning,* New York: Philosophical Library, pp. 353–365; Aronovici, C. (1944) 'Civic Art', in: Zucker, *New Architecture and City Planning,* pp. 366–391.
56. Lynch, K. (1960) *The Image of the City,* Cambridge, MA: MIT Press.
57. Sonne, W. (2009) 'Dwelling in the Metropolis: Reformed Urban Blocks 1890–1940 as a Model for the Sustainable Compact City', *Progress in Planning,* 72: 53–149.
58. Meitinger, K. (1946) *Das neue München. Vorschläge zum Wiederaufbau,* Munich: Munchner Verlag und Graphische Kunstanstalten, p. 7.
59. Sonne, W. (2006) 'Die Geburt der Städtebaugeschichte aus dem Geist der Multidisziplinarität', *Wolkenkuckucksheim – Cloud-Cuckoo-Land. Internationale Zeitschrift für Theorie und Wissenschaft der Architektur,* 10(2), available at <*www.cloud-cuckoo.net/*> (accessed 15 December 2011).
60. Hegemann and Peets, *The American Vitruvius,* p. 4.
61. Trystan Edwards, A. (1929) 'The Dead City', *The Architectural Review,* 66: 135–138; quote is from pp. 137–138.

PART II

Imagined townscapes

3

TOWNSCAPE AS A PROJECT AND STRATEGY OF CULTURAL CONTINUITY

Erdem Erten

> *. . . And to those for whom visual relations matter, the capacity to see represents itself as a way of salvation, just as for those whom social relations matter, forms of political arrangement represent themselves as a way of salvation . . . it is to architects that mankind will have to look to realize in visible terms a favourable environment for itself. It is thus the architect's role to become master coordinator, through whom the technicians of the statistical sciences, and the mechanical facts, as well as the painters and poets, must look to translate their raw material into the stuff of which visible civilization is made.*[1]

Introduction

Opening with a quote from Milton's *Paradise Lost*, the fiftieth-anniversary editorial manifesto of *The Architectural Review,* entitled "The Second Half Century," delineated the role of architects in post-war reconstruction.[2] Referring to the scene where Adam is about to see the future of humankind after his vision was cleansed, the quote sounds especially pertinent and poignant. The manifesto trusted and demanded from architects to coordinate the emergence of a new "visible civilization." It implied that in post-war reconstruction lay not only the hope of recovery from unprecedented destruction but a possibility of human salvation. For the leading periodical of the architectural community in Britain and arguably the rest of the English-speaking world, the creation of a favourable environment required the visual re-education of architects as well as the public, of their "capacity to see," a project of culture extending beyond the limits of architectural practice. The manifesto's concerns ran parallel to those of the state, as the decade that immediately followed the war was also seen by the state as a period of cultural as well as environmental reconstruction.

"The Second Half Century" is an intriguing, and maybe strange, document in the long history of the *Architectural Review* (*AR* hereafter) that stretches back to 1896. Foreseeing the drastic consequences of Britain's reconstruction, as well as investing it with great hope, the editors articulated a future course for the periodical aiming to shape post-war modern architecture. This course, formulated by Hubert de Cronin Hastings (chief editor of *AR* 1927–1973), J. M. Richards (joined as assistant editor in 1935, editor 1939–1970), and Nikolaus Pevsner (editor

1941–1970), was quite consistently followed in the editorial policies of the periodical throughout the quarter century (1946–1970) the trio led *AR*. During this period they published several series like "Townscape" (1949), "Outrage" (1955), "Counter-Attack" (1956), "Manplan" (1969) and "Civilia" (1972). These series assumed the nature of campaigns and were re-edited and published as books by the Architectural Press, such as Gordon Cullen's *Townscape* (1961) and Hastings's *Italian Townscape* (1963), not to mention several master plans for English cities by Thomas Sharp, Townscape's strongest advocate besides those directly affiliated with *AR*. In the form of campaigns intending to mobilize public opinion, Gordon Cullen, Ian Nairn, Kenneth Browne and Hastings (who wrote mainly under the pseudonyms Ivor de Wolfe or later de Wofle) had a particularly central role in disseminating "Townscape," which became the strongest discursive and pedagogical component of *AR* on shaping the city.

In an interview after Hastings's retirement from the periodical, Pevsner referred to Townscape as "policy."[3] According to him Hastings was a figure who preferred to remain in the background. However, being a proprietor of the Architectural Press and the chief editor of *AR* he certainly was the leading policy maker, bringing in the people for the campaigns or commissioning the books and special issues. A comprehensive reading of his essays, the books he wrote and his personal documents reveals that he had a more central role than any other editor who worked for *AR*. Townscape has usually been assessed by scholars as an urban design methodology in refusal of continental modernist urbanism with a resistance to theory, and it is criticized for nationalistic language, revivalism, nostalgia and class biases.[4] This chapter, however, will analyze Townscape as a project and strategy of cultural continuity, a strategy that stretched beyond the restricted domain of architecture and urban design, addressing the built environment as a whole.

A more thorough understanding requires seeing Townscape against the historical background that framed its conceptualization during and after the Second World War. In this period of growing disillusionment linked to global catastrophe and loss of territory, the looming threat attributed to pop culture from the other side of the Atlantic and the domination of technology, British thinkers on culture were preoccupied with a sense of cultural crisis. In parallel, the architectural community debated the viability of an international modernism and its planning principles supplied by Congrès Internationaux d'Architecture Moderne (CIAM). The field of cultural studies, in its developmental stages at this period, is characterized by its recognition of an alternative cultural vision in English popular resistance and Romanticism. Reinvesting these two traditions with contemporary meaning and a potential of recovery in order to revitalize British culture was key to the emergence and proliferation of Townscape, as it was in other fields of artistic and literary production. As confirmed by Hastings's personal documents the cultural policy that shaped Townscape was heavily inspired first and foremost by T. S. Eliot's *Notes towards the Definition of Culture* of 1948. After the Townscape campaign reached its peak in the late 1950s, Hastings found further confirmation in the writings of Raymond Williams, especially his seminal work *Culture and Society* (1958), both for his publication policies and for the advancement of Townscape's related campaigns.

During the post-war period culture freely travelled within different disciplinary fields, such as cultural studies, literature and anthropology, from which *AR* benefited.[5] In accepting the definition of "culture" as the whole way of life of a social group and seeing Townscape as its built counterpart, Townscape was complemented by articles on vernacular architecture, initially in sporadic articles, and after 1950 under the banner of "The Functional Tradition." The campaign for Townscape had a reiterative nature which the younger generation of architects bred in the rhetoric of avant-gardism found tedious. However, what makes the history of Townscape so intriguing is *AR*'s unswerving commitment to its continuity. What largely concerns this chapter is its relationship to models of culture and its early presentation as a strategy of "visual re-education," one that demanded to see the environment in a new light, which helps us to understand it as a project of cultural continuity.

T. S. Eliot's *Notes towards the Definition of Culture* and Hastings's editorial policies

According to the documents in the Thomas Sharp archive Hastings was already working on "a town planning theory which may materialize at one time or another" as early as December 1936.[6] This remark attests to the fact that he and others around *AR* were involved in the pre-war effort for what was first to be named Sharawaggi, and later Townscape.[7] Between 1942 and 1945, Pevsner and Hastings served as acting editors of *AR*. Town planning had assumed greater urgency during the war, and Hastings continued his search for the right author to materialize this project in the form of an operative text. He commissioned Pevsner to write a book on the history of the picturesque and its relationship to city development in Britain. The book aimed to clarify the English contribution to town planning in relation to British Romanticism and the landscape movement that ensued in consequence. In light of what is preserved in the Getty Institute's Pevsner Collection, the manuscript remained unpublished during Pevsner's and Hastings's lifetimes. Nonetheless, elements of Pevsner's research ended up as articles and helped in the commissioning of articles from other historians on the subject.[8] Some of this research would later resurface in Pevsner's Reith lectures for the BBC that would be compiled into his controversial but popular *The Englishness of English Art,* dedicated to Hastings.[9] However, Hastings's search for a book of similar content did not stop when Pevsner did not or could not provide it. During his collaboration with Sharp on the master plan of Oxford, published as *Oxford Replanned* (1948), Hastings thought that he had found the right author in Sharp, but his search would continue after Sharp did not provide the book in time.[10]

"The Second Half Century" announced that what looked like the sporadic appearance of the picturesque in *AR* that began in the early 1940s would be incorporated as policy in January 1947.[11] The editors stressed that increasing the visual quality of the environment was not simply a cosmetic activity, as the periodical's main objective was to recreate a visual culture. What laid the groundwork for the manifesto was a call for cultural change regarding the environment and a neo-romantic criticism of the Industrial Revolution.[12] What might have served as inspiration for Hastings and his launch of Townscape as an organized campaign, I will argue, is the publication of T. S. Eliot's seminal text in 1948.[13] Hastings possessed a copy of Eliot's *Notes,* and he carefully read, underlined and wrote comments on the book. His famous 1949 article, "Townscape: A Plea for an English Visual Philosophy Founded on the True Rock of Sir Uvedale Price,"[14] and the editorial policies that shaped *AR* in the post-war period carry strong links to Eliot's modelling of culture. At this time Hastings was also working on a manuscript titled "The Unnatural History of Man": his vision for a neo-romantic British society. Spurred by Hastings's concerns about the human condition after the nuclear catastrophe that ended the war, and about a possible role for the British in its future, the manuscript includes several references to theoretical positions on culture and cultural policy.[15] The following discussion aims to establish such links with specific reference to Townscape literature.

Townscape as culture and Townscape as cultural policy

Eliot's book was written at a time of early globalization, when international institutions such as the United Nations and its subordinate bodies like the UNESCO were writing their constitutions and grappling with the particularity of cultures within an international institution. This coincides with the period when CIAM had released the Athens Charter to set universal principles for modernist urban planning. According to Hastings, while Eliot refused to differentiate between culture and civilization, his model promoted the concept of cultural unity within diversity, and the necessity of particularization into regional cultures. Eliot defined culture as ways of life that people lived unconsciously, which he equated to practised religion. The understanding of sectarian variety as the differentiation of a central doctrine embodied a powerful model to explain cultural diversity, although Hastings crossed out Eliot's remarks linking culture to religion, probably in disapproval.[16] Eliot stated three important conditions for culture: (1) Culture is to have an organic structure to encourage the "hereditary transmission of culture within a culture," and this required the persistence of social classes; (2) it should be "analyzable, geographically into local cultures . . . [which brings up the problem of] regionalism"; and (3) "the balance of unity and diversity of religion – that is the universality of cult and devotion."[17] These three important conditions established a clear roadmap for Hastings in *AR*'s post-war history. Realizing that Eliot saw religion as a unifying force in European cultures Hastings noted on his copy:

> . . . the point surely is that a culture is differentiation at work as opposed to the identifying process of civilization i.e. city i.e. centripetal culture, so I make these the 2 kinds of culture, the differentiative principle which I call culture & the identifying which I call civilization, perhaps a civilization maybe the thing that unites different cultures . . . world culture should read world civilization . . .[18]

As Eliot saw a fertile conflict between regional cultures and class cultures, and the eradication of such conflict as detrimental to cultural change and evolution, Hastings believed that the dynamism of societies stemmed from the resistance of cultures to the unifying impact of civilization.[19] His stern resistance to the dominating models of CIAM and the garden city largely stemmed from the hostility of these models to existing urban contexts.

For Eliot, in order to ensure the transmission of culture the proper cultural policy would allow its evolutionary transformation without strict control. Culture could never be totally planned, but its transmission might be affected via education.[20] A total control over culture could stifle cultural change, and if necessary quasi-official bodies could be formed in support of cultural production and institutions. As Eliot stated: "For one thing to avoid is a universalised planning; one thing to ascertain is the limits of the plannable."[21] Therefore, wholesale conservation itself was not a productive strategy for Eliot. In his model the new should be in touch with old; that is the modern should always be in touch with the traditional. *AR*'s advocacy for the modern to establish links of continuity both in the single building and in the historical context of existing built environments answered Eliot's call.

Hastings's 1949 article is the official launch of Townscape under this banner and his most conscious attempt at promoting Townscape as a cultural attitude in shaping the city. In this article Hastings made another call for a reinterpretation of "picturesque theory" in order to generate "a functional vocabulary for the art of landscape" in the post-war city. The picturesque garden was a cultural precedent which was met with continental approval and as a result a "cultural best." This was effectively a search for the continuation of an aesthetic sensibility in combination with developments in modernist aesthetics, in order to avoid superficial historicism. For Hastings this would also secure the continuity of national character in the urban environment.[22] By opposing the planning ideas of the Modern Movement codified by the Athens Charter and differentiating "radical" picturesque theory – that is the work of Price and Knight against Repton's defence of Brownian landscape – from "debased" versions, Hastings argued that the picturesque was the "first Western radical aesthetic" that recognized the values of the "genius loci in a non-archetypal sense."[23] This argument could appeal to the anti-history avant-garde sensibilities of modern architects, and pointed to the potentials of the picturesque regarding place and culture.

The defence of the picturesque for Reyner Banham was "betrayal and abandonment" of the Modern Movement's ideals for the young generation of architects, for *AR* had established itself as the mouthpiece of the Modern Movement in Britain. Hastings, however, had cleverly directed his attack against modernist city planning for being in its "Lancelot Brown phase, with Corbusier, The Professor, busy about his clumps – clumps in the twentieth century of buildings rather than trees."[24] Equating Capability Brown's tendency of levelling the existing topography to create wide expanses of undulating grass surfaces to Le Corbusier's "will to demolition" as exemplified in the "Plan Voisin," Hastings pointed towards the dangers of the Athens Charter's universalizing tendencies and CIAM's direction as affected by Giedion and Le Corbusier. Planning as a cultural and social problem could not be overcome by technical procedures left to the operation of specialists detached from society.

A careful selection of cultural precedents followed Hastings's argument in order "to unite national character with the Spirit of the Age." Departing from empiricism as "resistance to theory" and as an attribute of Englishness, he constructed a mode of operation for what he called a "radical visual philosophy." The "townscaper" would follow yet another idiosyncratic English model, the Common Law, instead of relying on a theoretical framework. Based on the accumulation of cases that served as precedents and successively became part of law, the Common Law was hence open-ended, demanding critical reinterpretation with reference to the past. Compiled by Gordon Cullen and presented after Hastings's article was a "casebook." Defined by *AR* as "a mass of precedents gone over creatively to make a living idiom," it included urban design examples which the townscaper could learn from and reinterpret.[25] The creative reinterpretation of precedent was strategic in establishing continuity between the past, the present and the future.

The case-book aimed to demonstrate the logical consequences of the picturesque's application to the urban environment, such as multiple use, sudden shifts in scale, the value of urban enclosure in public spaces and the emergence of surprise. One of the major skills of the landscape "improver" of the eighteenth and nineteenth centuries was to evoke certain moods by means of the use of water, vistas, expansive prospects, ruins, enclosure created by means of trees and landforms etc., and to create diverse sensations in the observers as they moved through the landscape. The introduction stated that what followed was only a small part of a possible thesaurus that the visual planner could put together for himself, and that the case-book's main characteristics were its openness and incompleteness.

The case-book itself was based on a historical precedent, the pattern books, which allowed architects to publicize their work to clients and to demonstrate their skills and taste. Pevsner had written on the cultural effect of the pattern books in 1943 when *AR*'s campaign was evolving towards Sharawaggi.[26] The pattern book, by closely following the reigning taste of the period, was assumed to generate a unity of taste. For Pevsner, the pattern book was a source of commonsensical information that disseminated and unified architectural culture. As result of its popularity the developing middle classes were acculturated into aristocratic taste. By the end of the 1840s, the pattern books were replaced by architectural magazines and text-books that chiefly concerned architects. *AR*'s case-books would take on a similar role for architects and non-professional readers. Thus the readership, as well as the makers of the environment, would be drawn towards a consensus, and the cultural void left after the disappearance of pattern books would be filled.

At the end of 1949, after the launch of the campaign, the editors established "Townscape" as an unchanging section of the magazine, which would go on for about three decades. In the years between 1949 and 1958, when Cullen left *AR*, Townscape assumed the role of a laboratory of urban design. The pedagogy of Townscape was easily taken up by others who worked for *AR* and even by outsiders after Cullen left the magazine. In the meantime a majority of the Townscape cases handled in the magazine were actual urban design problems under the process of decision-making by local authorities and their planning departments.

New Marlow and the Thames linear national park in response to the new towns

Articles that focus on Townscape interventions which specifically emphasized cultural continuity illustrate how the discourse of Townscape was mobilized regarding real situations. After the New Towns Act (1946), the Town and Country Planning Act (1947), and the National Parks and Access to Countryside Act (1949), *AR* expanded the scope of its Townscape campaign to criticize and influence the implementation of the legislative framework. In July 1950 the plea to create a linear national park around the River Thames[27] was accompanied with Eric de Maré's largely visual analysis of the river via the categories of Patrick Geddes's "valley section" and his understanding of the place-work-folk trinity[28] (see Figure 3.1). He argued that the areas which surrounded the river could be seen as a region having developed a river culture. The national park could be created out of a regional survey of the landscape and the man-made components that made up this culture. Furthermore, de Maré argued that the river was once a major force of the economic life and the source of the culture that surrounded it, which made this area "the most English of English regions." The plea ended with the proposal of a new town around the area of Marlow, the design of which was based on the "river culture." After analyzing the visual qualities of the settlements and the natural beauty of the river in a case-book, *AR* proposed New Marlow as a social centre for the national park.[29] In Cullen's perspectives the town looked like a modern-day Venice (see Figure 3.2). A compact city without provision for growth, its centre was composed as a mixed-use, pedestrianized, high-density entertainment precinct served by canals and limited motor traffic. Surrounding the centre within a short walking distance were residential "basins," the equivalent of precincts in a college town.

Geddes had emphasized the need to transmit cultural heritage for a successful societal evolution through the "constant re-appropriation, selection, reinterpretation and reenactment of a fading past," a message that was shared by Eliot.[30] Although *AR* had referred to Geddes in the proposal,

FIGURE 3.1 "The Thames as a Linear National Park," special issue of *The Architectural Review,* July 1950

Geddes's idea of a comprehensive regional survey was not addressed. A major part of the issue was a visual survey captured by de Maré that zoomed in from the larger framework of the river's natural landscape to the minute details of the built environment. Geddes's idea of the survey, however, had a greater focus that involved a pedagogical program of civic education alongside scientific analysis that dealt largely with quantification of the environmental. While *AR* persistently condemned the planning profession for its obsession with the quantitative, it aimed to counter the quantitative by prioritizing the visual and consequently reinforcing an environmentally preservationist agenda. However, this preservationist attitude was far from promoting an eclectic and revivalist architecture that would generate a superficial continuity. In Cullen's images, all the buildings were consciously modernist, mostly alluding to the architectural vocabulary of Le Corbusier.

FIGURE 3.2 Gordon Cullen, perspective of New Marlow's town centre, *The Architectural Review*, July 1950

In the August 1950 issue, because of the 1949 reprinting of Geddes's *Cities in Evolution*, *AR* published an article by Lewis Mumford on Geddes, with an editorial introduction that presented *AR* as a follower of Geddes. His idea of conservative surgery, in contrast to Corbusier's radical surgery that followed Hausmann, aimed to recapture the lifestyle of a place and enhance it. Local and regional personality or, in *AR*'s favourite term, "character," had to be made an essential part of planning. Mumford also maintained that although Geddes respected major planning ideas of the twentieth century, such as the garden city and the neighbourhood unit, he kept a conscious distance from the constraints of a certain urban model, in this case either the garden city or the *ville radieuse,* like Townscape as policy did.

The connection between Townscape and culture came to the fore in another article in the same issue, where Cullen analyzed Brixham, Fowey, Looe and West Bay with reference to "*the line[s] of life,*" to demonstrate how the economics of the seaside and the natural geography affected urban form (see Figure 3.3). The lines of life were not simply formal features of the town but indications of its ways of life, its culture.[31] The townscaper's job for Cullen was to identify the lines of life in a city, analyze and categorize them in terms of importance, disentangle and reorder them if necessary and organize future development accordingly. If a town lacked coherence in the urban texture, the lines of life could be injected by future development in order to provide character. The powerful analogy established between the line of life and the way of life in the urban environment pointed to a complex relationship between culture and urban morphology while this complexity remained unaddressed in modernist planning schemes.

FIGURE 3.3 "The Line of Life," *The Architectural Review,* July 1950

Outrage, Counter-Attack and the Counter-Attack Bureau

Townscape cannot be fully understood without *AR*'s later campaigns, "Outrage" (June 1955) and "Counter-Attack" (December 1956), under the editorship of Ian Nairn. "Outrage" was published two years after J. M. Richards's article titled "Failure of the New Towns," in which he attacked the government's new towns policy for the insufficient commitment of capital and the lack of public buildings in the new towns to support a "culturally" full life. The new towns depended on London for the fulfilment of cultural needs,

FIGURE 3.4 Cover of the "Counter-Attack" special issue, *The Architectural Review,* December 1956

served as dormitory towns and nurtured sprawl and environmental dereliction. In June 1955 *AR* decided to enlarge its campaign to include the public and a larger environmental framework with "Outrage," expanding Richards's criticism and *AR*'s position within a special issue. The issue argued that subtopia, environmental blight encroaching upon Britain from the north to the south, was not a far prospect if the British landscape was left to change under misguided forces of development, when the ideal separation of town and country was violated.[32] *AR* aimed to influence citizens and local authorities to condemn subtopia and to mobilize the public for rehabilitation. The campaign's ultimate intention was to make a "sufficient [number of] people sufficiently angry." The issue ended with a call for action to the "layman" to stop his "country from being defaced," and it aimed to mobilize the whole population to claim the environment as cultural property. With a "checklist of malpractices," in closing the issue, *AR* equipped the layman to provide the necessary public surveillance.

"Counter-Attack" was planned to offer the public ways to approach the problems underlined by "Outrage," a remedy to cure subtopia (Figure 3.4). Following Townscape pedagogy, "Counter-Attack" collected precedents into another case-book. These precedents, however, hoped to illustrate how certain solutions succeeded in "bringing modern life to terms with the landscape" and "to arm the public against the wrong way and with examples of the right way of doing things."[33] Reflecting anti-American sentiment and targeting the rise of popular culture in Britain, the editors deemed Los Angeles the epitome of "planned" sprawl and a possible doomsday scenario for future Britain. Expanding J. M. Richards's earlier arguments, Walter Manthorpe held the whole planning machinery of the new towns responsible for subtopia through convincing comparisons of densities between the new towns and the desirable neighbourhoods of London. He recommended mixed-use development instead of functional zoning, urban redensification and adaptation instead of rebuilding, and flexibility in bye-laws and building regulations. Elizabeth Denby, the social reformer, condemned the decentralization policies of the government for planning against people according to pre-war data. The issue ended with *AR*'s call titled "A Plan for Planning," aimed to get support for a new planning attitude named "positive planning." The aim of positive planning was to "preserve and intensify the sense of place" and to develop an interdependent relationship between town and country to limit sprawl and to better allocate the resources of the countryside. "The only real guarantee of unspoilt countryside [was] a set of tightly planned towns."[34]

In June 1957 *AR* announced the opening of a "Counter-Attack Bureau" intended to serve as an advocacy service, to offer consultation and give voice to protest (Figure 3.5). Instead of applying to "Utilitarianism" and "Preservation" the bureau would adopt the planning attitude of "The Improver's."[35] The reference to "improvement" was by no means coincidental. In a later study Cullen argued that improvement was a vital strategy for the environment since the "new landlords" in the form of local or national authorities lacked the knowledge and love of a particular place.[36] One of the pedagogical objectives of Townscape was to help transfer the ideal gentry's cultural values to the citizen and to the planning authorities for the sake of cultural continuity. The model citizen and the notion of the state based on the gentry took a particularly formative role in Hastings's manuscript "The Unnatural History of Man" and its published version, the project for "The Alternative Society," which will be mentioned at the end of this chapter.

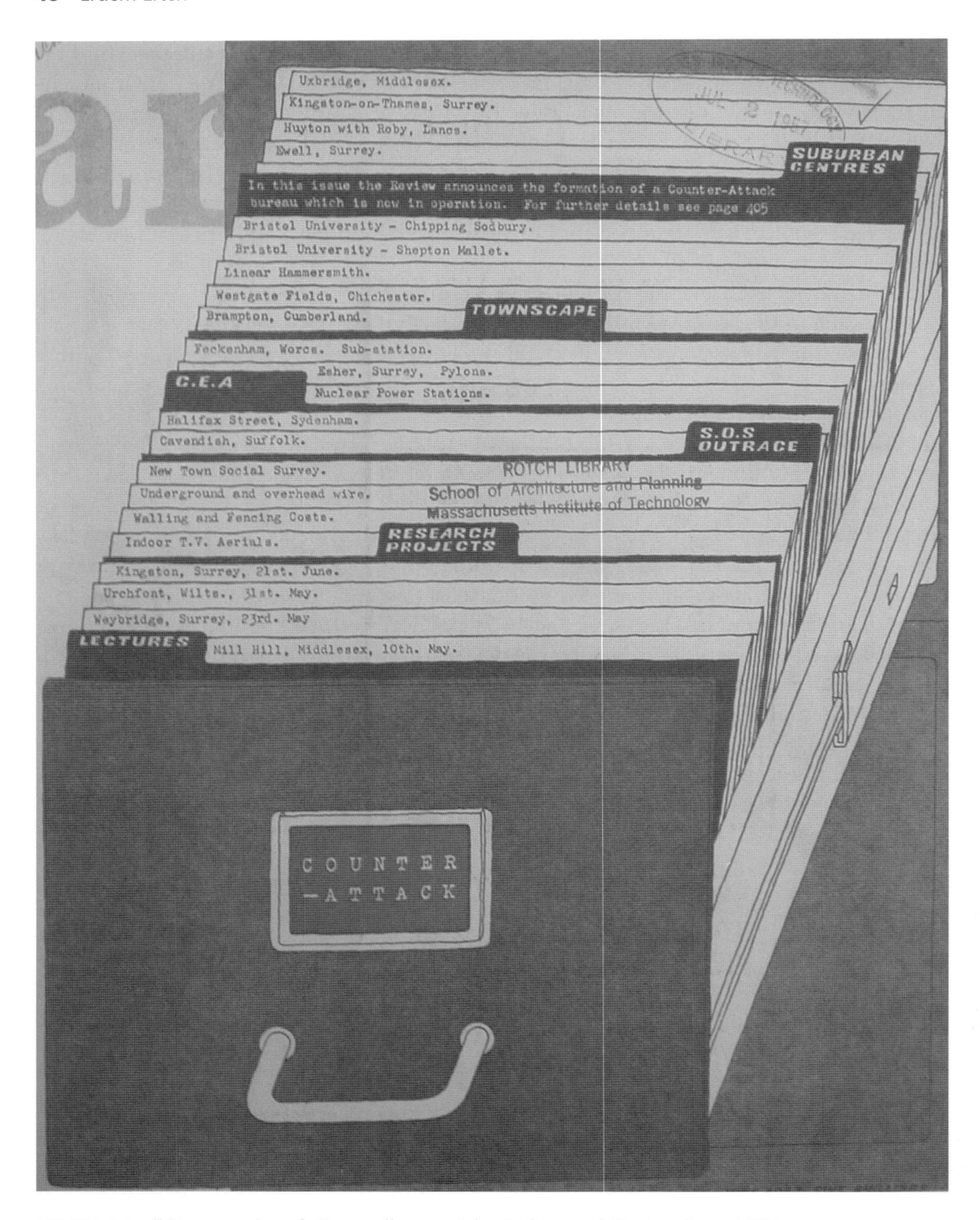

FIGURE 3.5 "Counter-Attack Bureau" cover, *The Architectural Review,* June 1957

Hastings's "alternative society" and a last call for reform

As Townscape studies continued to be published within *AR* in the 1950s and 1960s two important books which turned out to be canonical texts for the emerging field of urban design were released by the Architectural Press, with *Townscape* (1961) by Cullen and *The Italian Townscape* (1963) by

Ivor de Wolfe (Hastings's pseudonym). The former was a reworking of Cullen's studies published within *AR,* while the latter was largely a photographic essay on Italian hillside towns and an introduction in the form of a critique of modern planning practices. Many of the original textual introductions made for *AR* were omitted from Cullen's book, including Hastings's polemical 1949 essay that had launched Townscape. Among Hastings's papers is a draft outline for a book with the same name that he must have cancelled after having decided to concentrate on *The Italian Townscape* or presumably after Cullen went ahead with his *Townscape.* This draft partially served in the organization of the book.[37] In three chapters, the book opened with a revised version of Hastings's essay "The Theory of Contacts." Originally submitted to CIAM in 1937, the essay proposed to understand the city as a network of social and economic relationships, and the urban environment as the physical embodiment of such relationships.[38] Following this introduction Hastings aimed to criticize the garden city and the new towns as an absurd solution to town planning stemming from the popularity of the small garden, and to conclude his book with an analysis of the "European tradition" based on different senses of place in the town and its architectural counterpart such as the town wall as a point of transition, the town gate as invitation and the town square as arrival. In *The Italian Townscape* a similar analysis was given with added emphasis on the positive social role of congestion. Congestion was not an urban problem for Hastings, but a natural result of the human tendency to socialize, which needed to be celebrated via high densities. Hastings's emphasis on the social role of urban space is not at all addressed in critical work on Townscape.

Hastings's editorial interventions into *AR* were restricted to independent articles for a long time after *The Italian Townscape* until he started a new campaign in 1969 named "Manplan." Taking the opportunity to exploit popular culture's turn towards visual media and targeting the elections of June 1970 Hastings aimed to put "The Unnatural History of Man" into use, with editorial help coming from Tim Rock.[39] "Manplan" was almost a totally new magazine, interested in reformist policy-making and shocking the public by its provocative visuals with barely any Townscape content. Written in a rhetoric of "revolutionary humanism," the first issue aimed to voice "the sense of frustration" from which the editors believed that British society suffered (Figure 3.6). In the form of visual essays accompanied by captions, "Manplan" made a call for reform. In the editors' words "Manplan" meant "a plan for human beings with a destiny rather than figures in a table of statistics."[40] The issues starting from the second were thematically organized around communication (referring to transportation networks), industry, education, religion, healthcare and welfare, local government and finally housing. With each issue, "Manplan" deployed an attack on the above-mentioned fields and the inefficiencies of the capitalist consumer society. Politicians were asked to face the impact of consumer society's demands on the environment by drawing out a holistic, integrative structure prioritizing needs before profit. Arguing that industry had become less pollutant, "Manplan" reverted to the earlier arguments of Townscape to reintegrate the separated functions of the city for a more compact urban entity. The editors opposed the continuity of the new towns experience by arguing that industrial concentration proved to be wrong, expensive and socially segregating. In the fourth issue, on education, "Manplan" attacked the British public schools system as elitist and creating "a self-perpetuating oligarchy." In order to produce a society divested of class segregation, schools had to be integrated into the community and designed by user participation. In the early 1970s world that was becoming more and more suspicious of the objective truth of science the fifth issue of "Manplan," on religion, argued that religion could assume a new unifying role in order for society to deal with the consequences of industrialization. Viewed as a whole, "Manplan" should be regarded as an ecologically conscious and critical manifesto of the post-1960s consumer society in the aftermath of the Second World War.

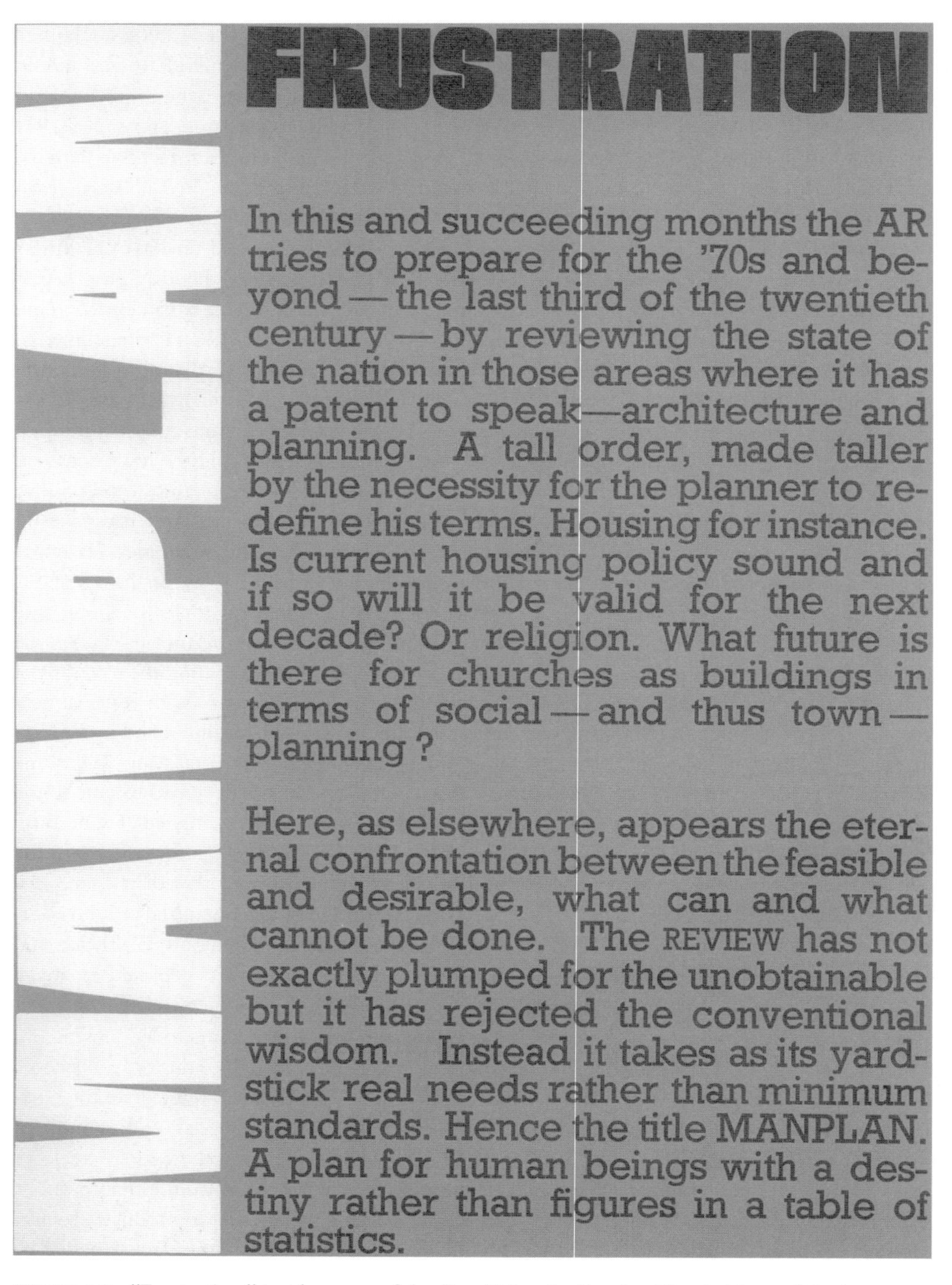

FIGURE 3.6 "Frustration," inside cover of the first "Manplan" series, *The Architectural Review,* September 1969

This two-year period brought forward an editorial crisis and a partial loss of revenue for *AR,* although the economic situation in Britain was partly responsible for the decline in advertising revenues. Even this did not discourage the campaigning spirit of Hastings. In 1971, with the help of his daughter Priscilla and of *AR*'s Townscape editors, Ian Nairn and Kenneth Browne, Hastings wanted to create in "Civilia" a new town that people would want to "belong to" before it was built. Under the pseudonym "Ivor de Wofle" he wrote an introductory essay titled "Towards a Philosophy of the Environment" that recalled the title of his essay for Townscape in 1949, enlarging the scope of his attention from Britain to the world. The basic task of the planner was to acknowledge cultural differences with reference to those of the regional and urban:

> . . . the planner's first priority is . . . accepting the culture of cities and the culture of regions as opposing goals – urbanism as international, unifying, centralising, consolidating, its apotheosis Marshall McLuhan's global village; regionalism as insular, microcosmic, separatist, differentiative, the culture of place. Only by accepting this opposition and acting in accordance with the difference between the two can a balanced society develop.[41]

Although "Townscape," "Outrage" and "Counter-Attack" were markedly against specialization and expertise, "Civilia" included a planning report.[42] The two planners approved Hastings's agendas by stating that twentieth-century planning policies had been threatening city centres. They recommended a "reversal of present, largely unplanned decentralisation trends by injecting new centres strategically placed" to attract sprawl and to rehabilitate existing centres.[43] Following environmental concerns Civilia was to be built on North Nuneaton's old quarries, a scar on the face of the land that was to be brought back to life with the healing hand of Townscape. "Civilia" was assembled out of photomontage and collage Townscapes juxtaposing images of Hastings's Italian Townscapes and well-known contemporary projects from the pages of *AR,* such as Moshe Safdie's Habitat, the viewing platforms of the South Bank exhibition and Paul Rudolph's Yale School of Architecture. This was a high-density, largely pedestrianized urban environment, built almost totally out of a New Brutalist vernacular.

Architect and planner Tom Hancock, the author of Peterborough's master plan, stated that if the authors of "Civilia" wanted social reform they should "describe a new city, an alternative city and by describing, imply a new society, an alternative society." *The Alternative Society: Software for the Nineteen-Eighties* (1980) was to be Hastings's last project, the only book published under his name and not his pseudonym. Heeding the calls of critics of "Civilia," Hastings published *The Alternative Society,* a reworking of "The Unnatural History of Man," as a policy of reform in Britain and around the world, which largely revealed the cultural project in the earlier post-war campaigns of *AR.*

According to Hastings, Britain could reinvigorate British traditions and animate them with a reformist spirit to give a utopian message to the world against the two prevailing political models: the savage capitalism represented by the US and the managerial and authoritarian socialism represented by the USSR. The new British society could be modelled on the political and cultural ideals of the "Cromwellian revolution" instead, which he thought was Britain's first real experiment with democracy.[44] Idealizing the protectorate, Hastings argued that the Cromwellian vision ought to be creatively reinterpreted in order to formulate a pluralistic democracy that would safeguard the welfare of every individual as well as that of the natural environment. In such a society scientific rationality and technological determinism would play only a secondary role in the human condition. The picturesque, "a radical, anarchist and disorderly ideal . . . [and] a tremendous event in the long apprenticeship of democracy," was to be the aesthetic metaphor of libertarian democracy, as Townscape was the physical manifestation of this ideal in the urban environment.[45]

Townscape a "creed": from the visual to the social

In April 1972, making the concluding remarks at the 2nd Dumbarton Oaks Colloquium on the History of Landscape Architecture, Nikolaus Pevsner bemoaned the state of city planning and urban design. He argued that the "missing link between the Picturesque and twentieth century architecture" was the English picturesque theory of the late eighteenth century. He ended his remarks by adding that he "always [tried] to preach the application of Visual Planning principles instead of the pernicious gibberish of sociological planning."[46] Pevsner's sentences stand out as affirmation of the *AR*'s editorial policy on town planning and urban design (which emerged as a new disciplinary field roughly during this period) between 1940 and 1974, when H. de C. Hastings, its chief editor for almost fifty years, retired.

From its very beginning in the 1940s to "Civilia" in 1971, "Townscape" aimed to reconcile modernity and tradition, and to create a resistance to capitalism by promoting a brand of Romanticism and portraying it as a natural continuity of national culture. While "Townscape" opposed modernist utopias in favour of piecemeal growth and the upholding of local cultures, "Outrage" and "Counter-Attack" voiced concerns for the environment and tried to mobilize the public on behalf of these concerns. "Civilia" united the aims of "Townscape" and the concerns of "Outrage" and "Counter-Attack" in a new town proposal by attempting to curb sprawl and to transform environmental dereliction into a model town. While the reformist intentions in *AR*'s proposals remained implicit in the formulating years of Townscape, partially in support of the young welfare state, during the following decades they became more and more explicit in opposition to the developing consumerism and consequent changes in British society. In spite of the very conscious and enduring nature of Townscape as a discourse, criticism mainly targeted its visual appeal and power. What the seductive and persuasive images concealed was a project of culture that remained to be deciphered.

Townscape still keeps its appeal in current urban design practice as an established pedagogical medium in creating and communicating the qualities of a desired urban environment. One major side effect of Townscape has been its encouragement of preservation and conservation, albeit that its emphasis on reconciling modernity and tradition seems to have been forgotten. By mobilizing architects, local authorities and communities for preservation and against development Townscape partially achieved its aim of creating a resistance. However, the broader environmental consequences of the pedagogy of "Outrage" and "Counter-Attack" have been limited against global economic development and the rise of consumer capitalism around the world.

Owing to its persistence within architecture journalism Townscape has been the most visibly successful among *AR*'s campaigns. Seemingly removed from the neo-romantic ideology that gave it birth, Gordon Cullen's *Townscape* is numbered among the canonical texts of today's urban design discourse and pedagogy. When "Collage City" was published as a special issue of *AR* in August 1975, *AR*'s editorial board turned their positions over to Colin Rowe, welcoming "Collage City" as a continuation of the "Townscape tradition." David Gosling's

(1934–2002) work in partnership with Cullen, and his teaching at Sheffield University and the University of Cincinnati, helped to disseminate Townscape as an urban design method. Townscape's focus on the urban context and its relationship to the community of inhabitants has had a major impact on larger-scale urban preservation. The urban clearance projects of the 1960s have given way to neighbourhood conservation and the acceptance of piecemeal urban intervention and high-density mixed-use development. As she would acknowledge later Jane Jacobs's work was highly influenced by Townscape and the American resistance to the anti-city visions of both CIAM and post-CIAM urbanism. Townscape's ideals have been largely incorporated into the agenda of New Urbanism.

"Counter-Attack" and "Outrage" have been followed as journalistic precedents for protests and have been taken up by periodicals around the world. Following the Counter-Attack Bureau, advocacy planning and community involvement in urban design have become standard democratic procedure in many places starting from the mid-1970s. *AR*'s pioneering interest in the relationship of architecture to the larger environment and its unremitting challenge to industrial development have found echoes in the discourse of sustainability and natural conservation.

In his Gold Medal Address to the RIBA in 1967, while he was still active on *AR*'s editorial board after twenty-five years with Hastings and Richards, Pevsner stated that what the historian could give to the architect above all was "a sense of continuity," summarizing the board's consensus.[47] Hastings's lifelong interest in Townscape reaching back to 1936 surely transcended an editorial mission. In Priscilla Hastings's words Townscape had become a "creed" for her father, a creed rooted in his version of British culture and in the possibility of an alternative society.[48]

According to Terry Eagleton what unites culture as utopian critique, culture as a way of life and culture as artistic creation is the failure of "culture as actual civilization" or "Culture" with a capital *C,* "as the grand narrative of human development." It might not be erroneous to state, then, that culture is bound to exist as resistance to civilization's failure, as culture is almost always theorized with reference to "its collapse, – in one sense the very definition of the concept becomes premised upon its decline."[49] Either in the form of the image of a sought-after future, or in the image of an ancient past that resembles an "emancipated future," culture is incessantly reformulated in interaction with the existing present. In this process culture gets transformed, reinterpreted, appropriated into forms other than the way it was actually lived, and that is how cultural continuity takes place. Townscape is such an attempt and a striking proof of this complex history.

Notes

1. The Editors, "The Second Half Century," *The Architectural Review,* 101 (1947), p. 24.
2. "Then Purg'd with euphrasy and rue / The visual nerve, for he had much to see." John Milton, *Paradise Lost,* book xi, line 414, in ibid., p. 21.
3. "We had a policy and went on with that policy . . . We had the policy of Townscape." Nikolaus Pevsner, interview transcript with S. Lasdun, 30 April 1975.
4. M. Bandini, "Some Architectural Approaches to Urban Form," in *Urban Landscapes: International Perspectives,* J.W.R. Whitehand and P.J. Larkham eds (London: Routledge, 1992), pp. 133–169; also see the Gordon Cullen Tribute in *Urban Design Quarterly* 52 (1994), pp. 15–30; N. Ellin, *Postmodern Urbanism* (Oxford: Blackwell, 1996); and A. Law, "English Townscape as Cultural and Symbolic Capital," in *Architectures: Modernism and After,* Andrew Ballantyne ed. (Oxford: Blackwell, 2004), ch. 8.
5. On this see M. Manganaro, *Culture 1922: The Emergence of a New Concept* (Princeton: Princeton University Press, 2002).

6. Letter dated 7 December 1936, from H. de C. Hastings to Thomas Sharp (University of Newcastle library, GB 186 THS 5.1.4).
7. For more on Hastings and Sharp's collaboration, and Sharp's involvement in Townscape's development, see E. Erten, "Thomas Sharp's Collaboration with H. de C. Hastings: The Formulation of Townscape as Urban Design Pedagogy," *Planning Perspectives,* 24,1 (2009), pp. 29–49.
8. According to the interview with Lasdun, Pevsner aimed to call the book *History of Visual Planning.* The incomplete manuscript at the Getty Center among the Nikolaus Pevsner papers was edited by Mathew Aitchison and published by the Getty Press under the title *Pevsner's Townscape: Visual Planning and the Picturesque* in 2010.
9. Nikolaus Pevsner, *The Englishness of English Art* (New York: Praeger, 1956).
10. See Erten, "Thomas Sharp's collaboration . . ."
11. The Editors, "The Second Half . . . ," p. 25.
12. Ibid., p. 31.
13. T. S. Eliot, *Notes towards the Definition of Culture* (London: Faber and Faber, 1948).
14. Ivor de Wolfe, "Townscape: A Plea for an English Visual Philosophy Founded on the True Rock of Sir Uvedale Price," *The Architectural Review,* 106 (1949), pp. 355–362.
15. The original manuscript is in the possession of Ms Priscilla Hastings, the daughter of Hubert de Cronin Hastings. The revised and shortened version was published as Hubert de Cronin Hastings, *The Alternative Society: Software for the Nineteen-Eighties* (London: David & Charles, 1980).
16. "Yet there is an aspect in which we can see a religion as the whole way of life of a people . . ." Eliot, *Notes . . . ,* p. 15. This sentence has a cross (x) next to it in Hastings's copy.
17. Ibid., p. 31. Hastings marked this part in his copy of the book as well.
18. This note is on p. 62 of Hastings's copy.
19. Ibid., p. 59.
20. Ibid., p. 94.
21. Ibid.
22. "Each age has its own priorities, and ours are social and technical, living as we do in an era of advanced scientific industrialization. But each age also has its constants; and of these by far the most potent is that temperamental bias of a whole community which under the name of national character continually demonstrates its enormous power of survival." De Wolfe, "Townscape . . . ," p. 361.
23. Ibid., p. 360.
24. Ibid.
25. Ibid., p. 362.
26. Peter F. R. Donner (pseudonym of Nikolaus Pevsner), "The End of the Pattern Books," *The Architectural Review,* 93 (1943), pp. 75–79.
27. "The Thames as a Linear National Park," special issue of *The Architectural Review,* 108 (1950). The issue was prepared by architect and photographer Eric de Maré (1910–2002). *AR* had also published a less detailed proposal similar to de Maré's in August 1949 by John Arrow, entitled "The [Norfolk] Broads as a National Park."
28. "[If the region] is to become a National Park this character, composed of Folk-Work-Place, to use Sir Patrick Geddes's well-known trinity, must be preserved and enhanced." De Maré, "The Thames . . . ," p. 7.
29. It is unclear whether New Marlow was intended as a pun on New Harlow, the first of the new towns that *AR* fiercely attacked due to lack of urbanity, since it was linked to Marlow.
30. Ibid., p. 41.
31. "The Line of Life," *The Architectural Review,* 108 (1950), pp. 95–106.
32. "Subtopia" was defined as "idealization of suburbia," that is a suburban utopia.
33. "Counter-Attack," special issue of *The Architectural Review,* 120 (December 1956), cover abstract.
34. Ibid., p. 431.
35. The editors argued that this attitude turned Barnsley, Gloucestershire, into a model village. Although the feudal property structure had dissolved in 1930 the village was bought back by the remaining members of the family to be saved from estate speculators. The houses were modernized and adapted to contemporary needs by an "improvement grant system." "Counter-Attack," p. 380.
36. Gordon Cullen, "Bingham's Melcombe," *The Architectural Review,* 120 (1956), pp. 100–104.
37. 1. **Theory of Contacts** / a. Town, a complex of contacts / b. Town planning and the resolution of those / c. Townscape and the art of interpreting that resolution in visual terms. 2. **The English interpretation** a. New Towns into Garden City / b. Absurdity of this solution / c. The fallacy of the small garden 3. **The European tradition** a. transition: the town wall / b. invitation: the town gate / c. introduction: narrow street / d. arrival: the town square / e. protection: the covered piazza.

38. See Reyner Banham, *The New Brutalism: Ethic or Aesthetic?* (London: Architectural Press, 1966), pp. 74–75. Also see on this Matthew Aitchison, *Visual Planning and Exterior Furnishing: A Critical History of the Early Townscape Movement, 1930 to 1949,* PhD dissertation, University of Queensland, pp. 230–234.
39. "Manplan" was published as four consecutive issues starting from September 1969, after which it became bi-monthly until the last issue appeared in September 1970, eight in total. Peter Davey, in his contribution to the special *AR* centennial number in 1996, notes that Hastings insisted on the change from the earlier layout of the journal into the form of visual essays. Richards opposed the idea and opted for the publication of special issues as before and for keeping the contents of the journal tailored to the existing readership. This opposition seems to have led to the wholesale renewal of the editorial board between 1971 and 1974, starting with the sacking of Richards from the position of executive editor, and Pevsner would soon leave.
40. "Manplan 1," *The Architectural Review,* 146 (1969), p. 173.
41. I. de Wofle ed., *Civilia: The End of Sub Urban Man, a Challenge to Semidetsia* (London: Architectural Press, 1971), p. 21.
42. The planners are listed as Rodney Carran (DipTP, AMTPI) and Michael Rowley (AADip, ARIBA)
43. Ibid., p. 27.
44. During the Civil Wars the English "believed it was the destiny of the English to bring in a New Order which would revolutionize the political, social and economic history of mankind." Hubert de Cronin Hastings, *Alternative Society,* p. 103.
45. Ibid., p. 103.
46. N. Pevsner, *The Picturesque Garden and Its Influence outside the British Isles* (Washington: Dumbarton Oaks, Trustees for Harvard University, 1974), pp. 120–121.
47. See Nikolaus Pevsner's Gold Medal Address in the *RIBA Journal,* 74 (1967), no. 8, pp. 316–318.
48. Unpublished note written by Priscilla Hastings in memory of her father given to the author in 2002.
49. Manganaro, *Culture 1922 . . .*, pp. 1–2.

4

VISUALIZING THE HISTORIC CITY: PLANNERS AND THE REPRESENTATION OF ITALY'S BUILT HERITAGE

Giovanni Astengo and Giancarlo De Carlo in Assisi and Urbino, 1950s–60s[1]

Filippo De Pieri

The historic townscape had a special relevance in the work of Italian architects in the decades following World War II, when many of the architects who considered themselves rooted in the 'new tradition' of rationalism (to borrow Sigfried Giedion's famous subtitle) searched for ways to adapt the conceptual framework of modern architecture to Italy's historic culture and environment.[2]

The impact of this attitude on the architectural debates of the 1950s and 1960s is well known, but some aspects of this story have been studied in less depth, most notably the influence that this attention to the historic townscape had on Italian planning instruments and practices, especially in the years following the adoption of the 1942 national planning law.

Two plans by Giovanni Astengo and Giancarlo De Carlo, respectively, drafted between the late 1950s and the early 1960s for the cities of Assisi and Urbino, are interesting and influential examples of the way in which postwar Italian planners dealt with the preservation of historic cities. Although never implemented, the plan for Assisi became well known nationwide, thanks to its early publication and the prominent role played by its author Astengo in Italian planning circles.[3] The plan for Urbino received considerable attention outside Italy and was the only Italian planning document of the time to enjoy a full translation into English.[4]

Although Astengo's and De Carlo's plans for Assisi and Urbino cannot be taken to represent Italian planning culture as a whole, they do provide significant examples of the way in which the historic city was analysed and represented by Italian planners after World War II. The two plans were ambitious documents that aimed to provide an application of state-of-the-art planning techniques. Nevertheless, when confronted with what were both small, declining medieval cities in central Italy, Astengo and De Carlo faced some difficult and, in many ways, comparable problems. The historic built and human environment of these cities seemed to pose challenges to which planning tools grounded in CIAM (Congrès Internationaux d'Architecture Moderne) and post-CIAM traditions did not appear entirely suited. Hence both plans were characterized by methodological eclecticism, an approach possibly more indebted to European traditions of visual or picturesque planning than their authors were willing to admit.

Modernizing Assisi and Urbino

A few years separate the plans for Assisi and Urbino: Giovanni Astengo (1915–90) started work on his plan for Assisi in 1955, and the plan was adopted by the city council in 1958. Giancarlo De Carlo (1919–2005) started work on the Urbino plan in 1958, and it was adopted by the city council in 1964.

In 1954, Assisi was included by the Ministry of Public Works in the list of the first 100 Italian cities that had to approve a general planning document, in line with the prescriptions of the national planning law of 1942. Astengo was already recognized as a leading personality in Italian planning, particularly because of his editorial work for *Urbanistica,* the journal of Italy's Town Planning Institute.[5] Nevertheless, the appointment was his first professional opportunity to take responsibility for the plan of a city (Fig. 4.1). Astengo did not hesitate to invest time and money in the plan, renting a house in Assisi (his professional base was in Turin) and bringing with him several collaborators. These efforts, however, ended in failure, both economically and professionally. Changes in the city's administration induced the city council, initially supportive of the plan, to consider it with increasing suspicion. The highly restrictive regulations concerning new residential or industrial areas caused growing opposition among local actors. A 'special law' for Assisi approved by the Italian Parliament in 1957 – a law that allocated special funds for urgent repairs and new investments in the city – interfered with Astengo's work in a way that turned out to be decisive. The city council finally decided to abandon Astengo's plan in 1959.[6]

De Carlo's architectural involvement with Urbino began in the early 1950s. Soon after his firm opened in Milan, he obtained a few architectural commissions from the University of Urbino with the support of its influential dean, the literary critic and historian Carlo Bo. His work in the city was to continue intermittently over the following 50 years, up to his death in 2005, turning Urbino into one of the key sites of his lifetime career.[7] Some of his architectural interventions in the city have subsequently been read by critics as implementations and developments of ideas expressed in the 1964 plan.[8]

In 1958, Urbino had also been ordered by the Ministry of Public Works to draw up a general plan for the city. De Carlo was not immediately commissioned for the task but was asked to coordinate a few preliminary studies that ultimately led to his being appointed in 1960 as the consultant responsible for the plan (Fig. 4.2).[9] The plan was commissioned by the city council of Urbino, led by its Communist mayor Egidio Mascioli, but the influence exerted by the university on the city was decisive in the choice of De Carlo. Of particular importance was the role of the philosopher Livio Sichirollo, a professor at the University of Urbino who served as deputy mayor for planning between 1956 and 1970, acting as the architect-planner's local reference point and one of the masterminds behind the plan.[10] At the time when the plan was in preparation, De Carlo was also involved in an attempt to draft a metropolitan Master Plan for the Milan region. Entirely different in their scale, these two undertakings from the 1960s marked a period in De Carlo's career when his interest in planning issues was possibly at its strongest.[11]

For both architect-planners, their work on the plans for Assisi and Urbino was part of a longer involvement with the local context. The writings accompanying the plans convey the sense of an almost anthropological experience in the careful way that the architect-planners explored the small cities they had been called upon to re-organize. Small cities such as Assisi and Urbino (which had populations of about 20,000 at the time of the plans) were a type of

FIGURE 4.1 Assisi. Cover of the special issue of *Urbanistica* dedicated to the plan, pp. 24–25, 1958 (courtesy Inu Edizioni, Rome)

FIGURE 4.2 Urbino. Cover of the English edition of Giancarlo De Carlo, *Urbino* (MIT Press, 1970)

setting unfamiliar to these architect-planners, who had hitherto been involved in undertakings on a larger scale. Both De Carlo and Astengo conducted extensive research on the economic and social structures of the cities and on their housing conditions. Indeed, Astengo's preliminary survey of Assisi was so consistent and detailed that his plan was seen by many as a turning point in Italian planning practice.[12] The two planners believed that it was necessary to cultivate a long personal acquaintance with a city in order to reach a full understanding of its social and built environment. In their view, one of the tasks of the planner was to identify the character of the city and its individual traits, an identity from which the plan had to take inspiration in order to adapt its tools to those material and immaterial features that contributed to the uniqueness of a place.

Two different planning strategies

Both Astengo in Assisi and De Carlo in Urbino saw their plan as an opportunity to suggest an ambitious socio-economic strategy – a task that went far beyond the scope of the documents which they had been charged to draft. Each believed that their city was undergoing serious decline and saw their plan as the potential catalyst for a new phase of economic and social planning, taking the opportunity to propose a number of directions this might take. Nevertheless, their recommendations for Assisi and Urbino were quite divergent.

Astengo saw an increase in agricultural production and the initiation of activities in the secondary and tertiary sectors as the key measures to be pursued in countering Assisi's 'economic depression' – as he called it – and socio-demographic crisis. The purpose of his plan was to prepare the ground for these developments, particularly through the regulation of land uses and the provision of infrastructure, while ensuring the most careful preservation of the historic built environment. The plan did not rule out the risk of further demographic and economic decline but deemed it possible to contribute to reducing, and to partly reverse, these trends in the long term. It proposed locating most of the new economic activities outside of the historic city, regarding the latter as a collectively built object that demanded meticulous protection.[13]

De Carlo's plan focused less on production and more on culture. It established a link between the future development of Urbino and the evolution of two trends that were thought to be already in place: the growth of the university and the growth of tourism. The historic centre was envisaged by the plan as the strategic locus that could guide Urbino's 'general recovery'. Interventions in the historic city and in the surrounding landscape were to be closely connected and part of a common strategy – as the choices made by the university for its buildings both inside and outwith the walled city were already making clear. At the time of the plan, tourism in the region was mostly directed to the leisure centres of the Adriatic Coast (such as Rimini or Riccione), but De Carlo believed that it was possible to promote 'a more equitable distribution of the resources' so that 'the tourist potential of the whole area' could be 'exploited to the advantage of all concerned'. Although he was highly critical of the way in which mass tourism induced a superficial consumption of urban space, he also believed, quite optimistically perhaps, that 'brighter prospects would be possible for the future, in which mass tourism will have superseded its present impetuous and barbarous stage of conquest – a future when not just amusement but culture is sought as well'.[14]

The Assisi plan was based on the definition of strict regulations and zoning prescriptions: Astengo firmly believed in the governing role of public institutions, and his plan aimed to provide them with a tool that could help them control the choices of private actors, while serving as a coordinating scheme for other planning policies. De Carlo was more suspicious about zoning and prescriptive regulations,[15] and sceptical about the idea that a city could be ruled largely by means of public authority. His plan was not presented as a document requiring full implementation but as a set of possible strategies, to be pursued by both public and non-public actors. It was described as a 'perspective plan'.[16]

Both plans paid great attention to the design of road networks surrounding the city and to the issues raised by road transport, and proposed almost opposite solutions to the problem of automobile circulation. De Carlo recommended the pedestrianization of the historic city as a whole, since 'no attempt to regulate the situation will ever successfully marry this city, with its typical forms, to a mode of transport quite alien to its very nature'. Astengo proposed rationalizing automobile circulation (Fig. 4.3) and removing surface parking from the main squares, but he did not consider pedestrianization as a solution: 'it would be anti-historical to bring today's cities, even when they are charged of ancient documents, back to the situation of a few centuries ago'. Both plans included detailed architectural designs for new car parks to be constructed immediately outside the city walls.[17]

Despite their different approaches, the two plans shared a strong focus on all matters related to the protection and conservation of the historic townscape. They were extremely restrictive in

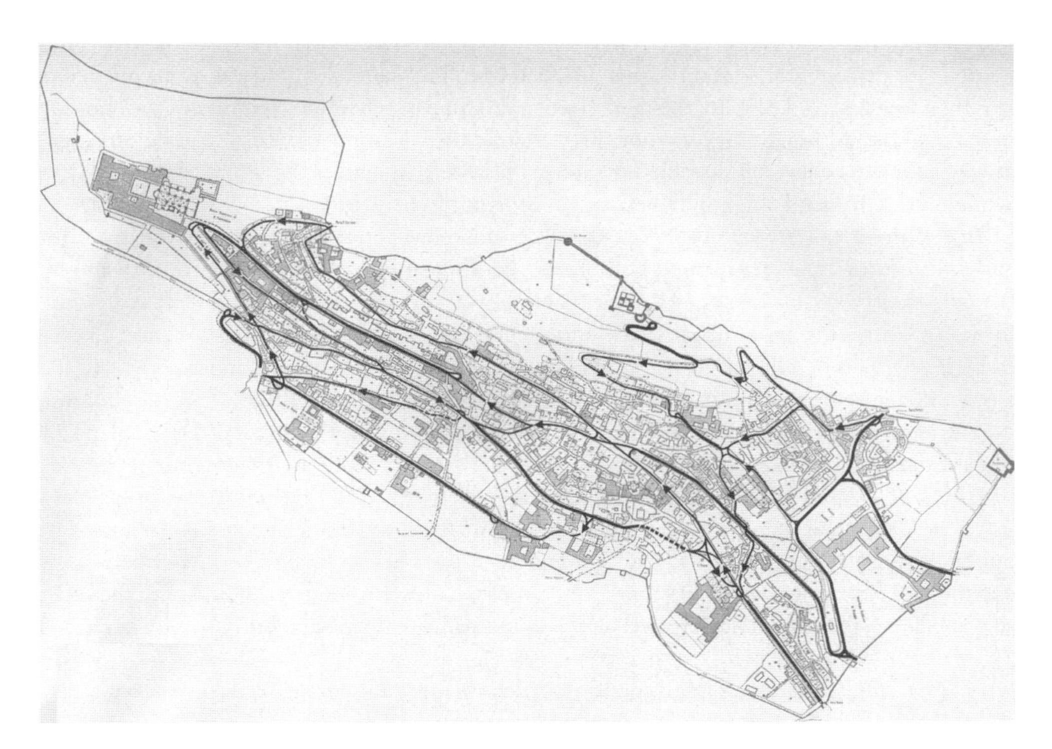

FIGURE 4.3 Assisi. New organization of automobile traffic in the historic centre, as proposed by Astengo's plan. From *Urbanistica,* 24–25, 1958, p. 105 (courtesy Inu Edizioni, Rome)

the provision and location of land for new buildings and insisted that, in order to preserve historic cities like Assisi or Urbino, a general plan should be complemented by other, more detailed planning instruments that needed to be elaborated simultaneously with the general framework. An important component of the two works was dedicated to a study of the historic city, both in its present-day form and in its past evolution. Neither De Carlo nor Astengo engaged historical consultants for this purpose, much in the same way that they did not employ specialist consultants on economic analyses or transport studies. In a sense, their plans were good examples of the 20th-century Italian tradition that considered architect-planners, with their body of technical and non-technical knowledge, as fully capable of dealing with the complex factors shaping urban change.[18]

It is now appropriate to examine more closely the conceptual and visual tools used by these two architect-planners in order to understand the historic townscapes of Assisi and Urbino and to represent them in ways that could positively affect the planning process.

Narratives of urban evolution and decline

The historical presentations that accompanied the two plans read the history of the two cities within a broad developmental framework. Both plans suggested that an initial, formative period in the city's history had been followed by a phase of culmination in which the key elements of the historic urban structure had taken shape. The cities had then faced a long phase of relative stagnation, and finally a steady decline that coincided with more recent transformations, such as those of the late 19th and 20th centuries. At the peak of their development, the built environments of Assisi and Urbino appeared to have reached an internal coherence and complexity that subsequent transformations had been unable to sustain. Both plans projected, and therefore had to deal with, the memory of a distant past when the city formed a seemingly organic whole. Although they did not aim to restore the harmonic traits of the urban structure at its peak, they believed it possible for modern planning to respect the traces of this heritage, while putting forward an entirely new urban organization.

In Astengo's view, the pinnacle of Assisi's development could be found between the 13th and 14th centuries, an era marked both by the communal liberty enjoyed by the city and by the personality of Saint Francis. The spiritual upheaval and 'great fervor of life' inspired by the Saint, especially after his death in 1226, left indelible traces on the townscape of Assisi: 'Everything here speaks of the Saint, and a sweet serenity, sustained by awareness and energy, is in the air'.[19] Such an emphasis on the influence of Saint Francis was interesting, especially because it contrasted with the fact that the plan did not seem to consider the presence of pilgrims or religious institutions in Assisi as a factor in potential economic or cultural development.

Astengo analysed at some length the structural elements of the medieval city: in his view, it was especially the way the city was embedded in the natural landscape and the shape of its public spaces that contributed to the uniqueness of the place. The city was described 'as an ensemble of *piazze* situated at different levels', all of which were 'rectangular and with the main axis parallel to the mountain', crossed by roads and paths that repeated cruciform patterns.

> The same structural system, repeated in the major and minor squares, gives the city an exceptional unity that can immediately and instinctively be perceived, although an uncommon effort of abstraction and synthesis is needed in order to fully understand it. Such a system enriches the city with an extraordinary variety of effects, [. . .] so that, for each monument, all points of view become possible at once [. . .]. The simultaneous

> structural unity and multiplicity of urban spaces, added to the distant views of the mountain and the plains, that [. . .] appear in the built landscape, re-affirming the constant presence of the natural support, forms the great richness of Assisi's urban setting.[20]

The decline and physical dilapidation of Assisi's urban structure started in the 15th century, and important alterations came in the 17th and 18th centuries, when, for example, the construction of large baroque palaces brought serious modifications to the scale of the city. Nevertheless, Astengo maintained that these did not seem to radically compromise the possibility of reading the city as a whole. Things changed after Italy's unification, however, a period that Astengo identified with the 'recent ruin' of Assisi. Two factors contributed to the steady decline of the built environment: the pressures imposed by modern infrastructure (starting with the construction of the railway line in 1870) and the inadequate – if not wholly disruptive – insertion of recent architecture into the townscape. In the architect-planner's words, the late 19th and 20th centuries had been a period of 'confusion of ideas and vandalistic destruction'.[21]

The highest point of Urbino's history, according to De Carlo, could be situated at a somewhat later date, during the years of the early Renaissance and the rule of Federico da Montefeltro (1422–82). The plan placed particular emphasis on the modifications brought about by Federico and his architects, centring on the work on the Ducal Palace. The re-orientation of the palace, with the construction of its new 15th-century façade, seemed to have radically altered the spatial structure of the city in both functional and symbolic terms:

> The volumes of the Palace swing around from the broad channel of the piazza to the line of the ducal apartments. During this rotation, the architectural language also changes from the precise and functional nature of the portions flanking the city center to the nearly utopian impact of the Torricini, rising like banners over the landscape. This movement is reflected and amplified by the surrounding pattern. It produces a well-knit sequence of spaces that descends like a spiral from the summit of the first hill to the Abbondanza donjon, crossing in front of San Francesco on the way down the slope of Lavagine and the ends of the two streets that descend from the Monte and Santa Lucia wards.[22]

Once again, interventions from the 16th century onwards only managed to alter what was interpreted as the nearly perfect urban balance reached during the age of Federico. Like Astengo, De Carlo was highly critical of contemporary attempts to modify the urban fabric: most city planning subsequent to Italy's unification seemed to him to be marked by 'incompetence and stupidity'.[23] Nevertheless, his overall judgment was more mixed: not all modern interventions were to be dismissed, and some could be considered appropriate enough to the scale and structure of the city. This was the case especially with the construction of the Sanzio theatre, designed by the architect Vincenzo Ghinelli during the first half of the 19th century. De Carlo was also keen to stress that even the most disruptive transformations of the last few decades had their roots in specific social demands, to which they had attempted to respond. He argued that Urbino had suffered a serious decline, especially due to the decreasing role played by the city at regional level. The unification of Italy had inflicted calamitous impacts on the city because of the decision to link Pesaro with Urbino as the joint provincial capitals and because of the lack of infrastructural modernization.

By proposing historical narratives that condensed the ultimate sense of the history of a city in a particular phase of its development, Astengo's and De Carlo's plans were influenced by traditions of urban analysis that dated back to the previous century, if not further. It is tempting to perceive a distant echo of Camillo Sitte's influence in the fact that a relevant section of the two plans was dedicated to the detailed study of an idealized and nearly perfect stage of each city's spatial evolution, which seemed to contain many lessons for the modern planner. But the narratives proposed by the plans also had deep local roots and were indebted to the work of 19th- and early 20th-century memorialists and local historians.[24] Finally, the evolutionary approach to urban history privileged by the two architect-planners helped them to build an important argument for the political legitimation of their plans. A symbolic connection was implicitly established between the urban golden age of the medieval period or the early Renaissance and the new urban phase envisaged by the new plans. The two documents were presented as the historical opportunity to rediscover the identity of each city and re-affirm it in the context of Italy's economic modernization. Astengo went so far as to declare that a plan for a city could be considered as 'a modern version of the ancient communal statutes'.[25]

Visualizing the historic city

The plans for both Assisi and Urbino made extensive recourse to visual materials for their analysis of the historic city. Visual analysis was instrumental in allowing the architect-planners to communicate messages that went beyond the purely technical scope of their plans: it evoked the persistence of an urban community whose fundamental unity had somehow resisted the uncertainties of history. Nevertheless, visual materials were used by the two architect-planners in significantly different ways.

FIGURE 4.4 Assisi. View of the city. From *Urbanistica,* 24–25, 1958, p. 14 (courtesy Inu Edizioni, Rome)

Photography had an especially important role in Astengo's survey of Assisi. His papers document the painstaking search for different kinds of visual sources, and those that illustrated the plan represented only a fraction of the impressive wealth of material accumulated during his work.[26] Such an interest was consistent with Astengo's implicit belief that the essential characteristics of the historic city were those that could be *viewed*.[27] His approach was grounded in an attitude towards historic cities that tended to appreciate their pictorial values.[28] The photographs published by Astengo attempted to document an abundance of traits relating to the city's visual character: its general topography, its architectural details, its streets, its visual relationships with the countryside (Figs 4.4, 4.5). Images were mostly in black and white, but colour photographs were also deemed important due to the chromatic variations offered by the architecture of the historic city.

The photographic surveys aimed at providing a detailed understanding of all aspects of Assisi's anonymous architecture. In Astengo's view, his understanding needed to be 'critical', that is, based on a correct judgment of the historic city's architectural 'values', so as to distinguish the essential elements of the built environment from the disturbing alterations imposed

FIGURE 4.5 Assisi. 'Authentic medieval elements'. From *Urbanistica*, 24–25, 1958, p. 111 (courtesy Inu Edizioni, Rome)

on the townscape.[29] Such an approach inspired one of the most spectacular pieces of visual analysis in Astengo's presentation, a collection of manipulated photographs presented under the title *Assisi's recent ruin*. The series consisted of black-and-white pictures of Assisi's built heritage where recent developments that had dramatically altered the visual balance of the city were emphasized by means of a yellow overlay. Astengo was especially keen to emphasize those 19th- and 20th-century architectural interventions that had adopted historical styles, thus contributing, in his view, to shaping a fake medieval townscape that perturbed the 'original' townscape of Assisi.[30]

There were many implications in such a presentation. Astengo depicted historic Assisi as a single, unique human artefact, formed slowly through the centuries, to which only the last century had introduced dramatic alterations. Although the plan did not propose to demolish the contrasting architectural elements, the reader was implicitly invited to imagine the landscape of Assisi without them. In order to restore the visual unity of the historic townscape ('The face of Assisi', as the opening pages of the presentation in *Urbanistica* called it),[31] it seemed important progressively to carry out specific actions on each of the buildings and the architectural details that composed it. The conservation of townscape was, above all, conceived to be a problem of architectural restoration.[32]

De Carlo did not share Astengo's interest in the visual analysis of architectural elements, nor did he consider the restoration of buildings to be the central aspect in a modern approach to the historic city. His photographic material did not attempt to provide an exhaustive documentation of the physical characteristics of the urban fabric: on the contrary, it was largely concerned with a few parts of the city (most notably those surrounding the Ducal Palace) (Fig. 4.6).[33] Such an approach reflected an interest in the city's public spaces and the role played in them by monuments. De Carlo saw the history of Urbino as being particularly marked by the changing relationships between monuments and urban space. This position could be seen as a modern version of the traditional interpretation of Urbino as a city shaped by its principal buildings: a 'city in the form of a palace', as Baldesar Castiglione had famously stated in his *Book of the Courtier* (1528).[34] On the other hand, it resonated with several lines of recent architectural research, especially those opened by the postwar CIAM debates about such issues as 'the new monumentality' and the 'heart of the city',[35] which were to be continued, with a different accent, in Aldo Rossi's *Architecture of the City*.[36]

A few general remarks, formulated by De Carlo in his book on Urbino and in other writings of the time, suggested that understanding the character of a historic city required a strong effort of intellectual abstraction. 'Structures' and 'forms' were two recurrent keywords – the former being 'the organizational systems that allow the activities of individuals and social groups to develop in space', the latter the 'physical materialization' of such structures. Forms could, in a few special cases, prove particularly 'resistant to change' and embody a 'utopian charge' capable of stimulating a further transformation of the environment. The settings (*ambienti*) resulting from this interplay of forms and structures comprised both the built and the natural environment. The task of the planner in a historic city was to decipher the changing configuration of structures and forms over time. The central problem was the 'selection and classification' of 'ancient forms' on the grounds of their adaptability 'to the new organizational patterns planned to meet the city's contemporary functions, as well as those of its hinterland'.[37]

FIGURE 4.6 Urbino. Aerial view. From De Carlo, *Urbino*, p. 88 (courtesy IUAV, Archivio Progetti, Fondo Giancarlo De Carlo)

Photographs or historical maps and views were clearly insufficient to convey such an analysis, and, in fact, De Carlo's plan of Urbino made extensive recourse to diagrams and graphic schemes. These illustrated every aspect of the plan and were an interesting attempt to present the results of the analytical work in a way that could be useful for the future design of the new Urbino.[38] The graphic design of the 1966 book, conceived by Albe Steiner, gave these materials great weight.[39] Some of De Carlo's diagrams evoked – both in their titles and in their graphic layout – Kevin Lynch's schemes for *The Image of the City* (1960), most notably the two diagrams dedicated to the 'visual analysis of the town' and the 'visual analysis of the landscape' (Fig. 4.7).[40]

A few diagrams were especially developed by De Carlo for his 'historical' analysis, in order to provide a visual translation of the shaping factors behind the city's transformation over the centuries. The history of the city was articulated into four historical phases ('The Roman town', 'The medieval town', 'The Renaissance town', 'The nineteenth-century town') (Fig. 4.8): for each phase, by means of various graphic symbols, the drawings singled out a few elements within the urban organization, such as the orientation of the main axis of circulation, the points of access to the walled city and the concentration of civic functions. A fifth diagram

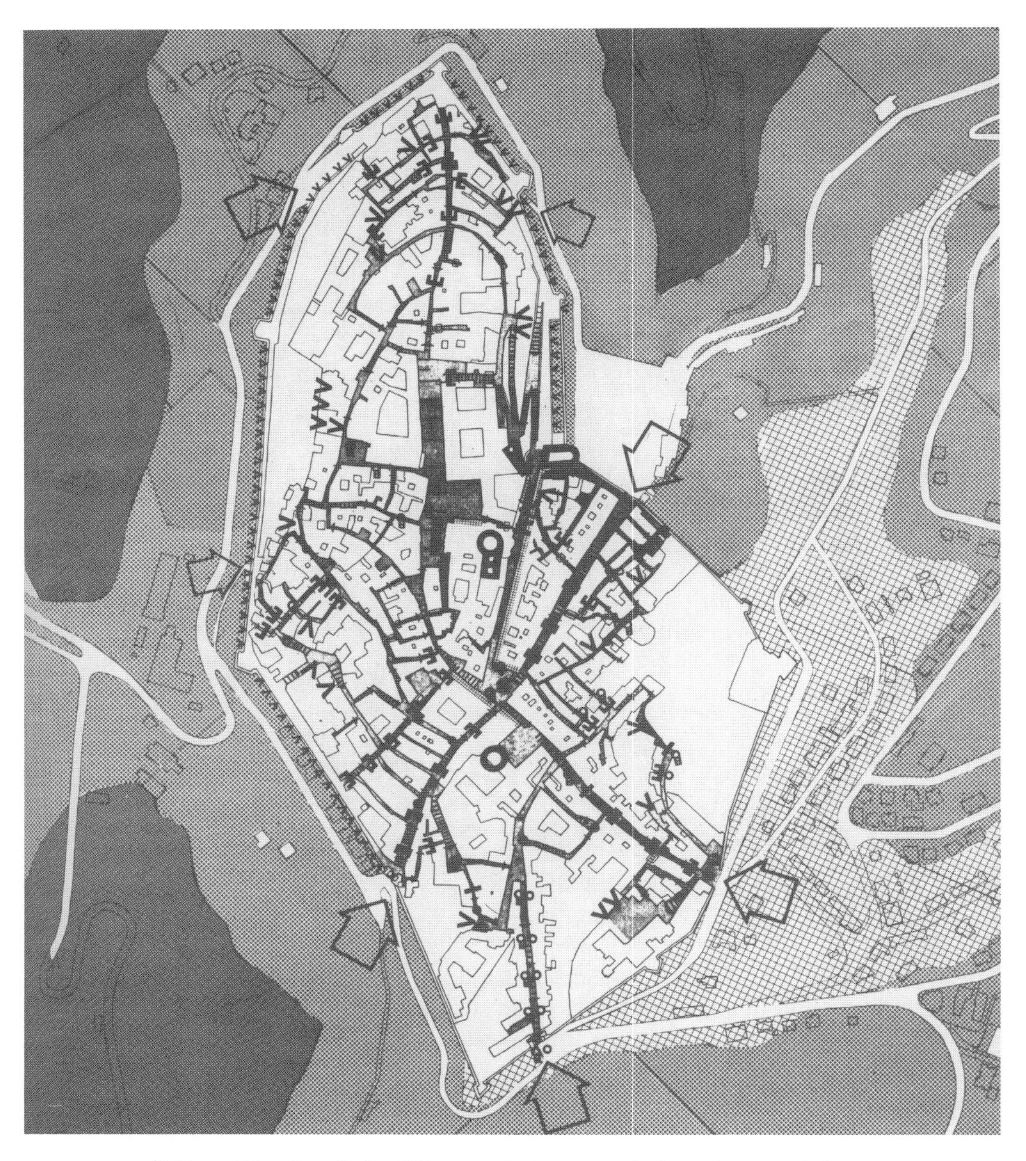

FIGURE 4.7 Urbino. 'Visual analysis of the town'. From De Carlo, *Urbino*, p. 71 (courtesy IUAV, Archivio Progetti, Fondo Giancarlo De Carlo)

('The city in the future') closed the sequence with the presentation of the key actions of the plan. Taken as a whole, the series seemed to imply that the history of Urbino could be broken down into a sequence of separate moments, each with its own distinctive traits. Seen from this perspective, the plan did not appear different in nature from the great collective forces that had influenced past urban transformations.[41]

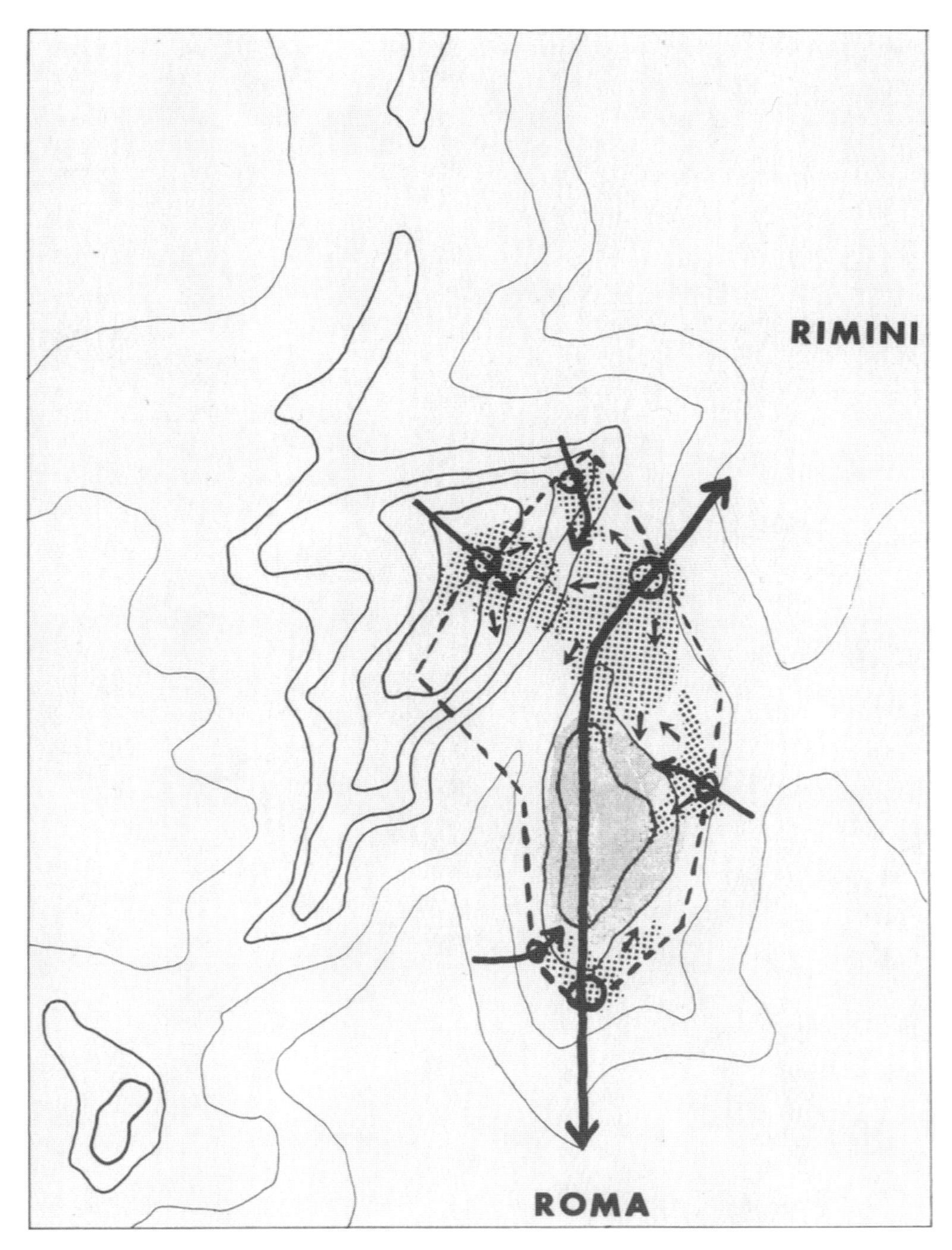

FIGURE 4.8a Urbino. 'The medieval town'. From De Carlo, *Urbino*, p. 98 (courtesy IUAV, Archivio Progetti, Fondo Giancarlo De Carlo)

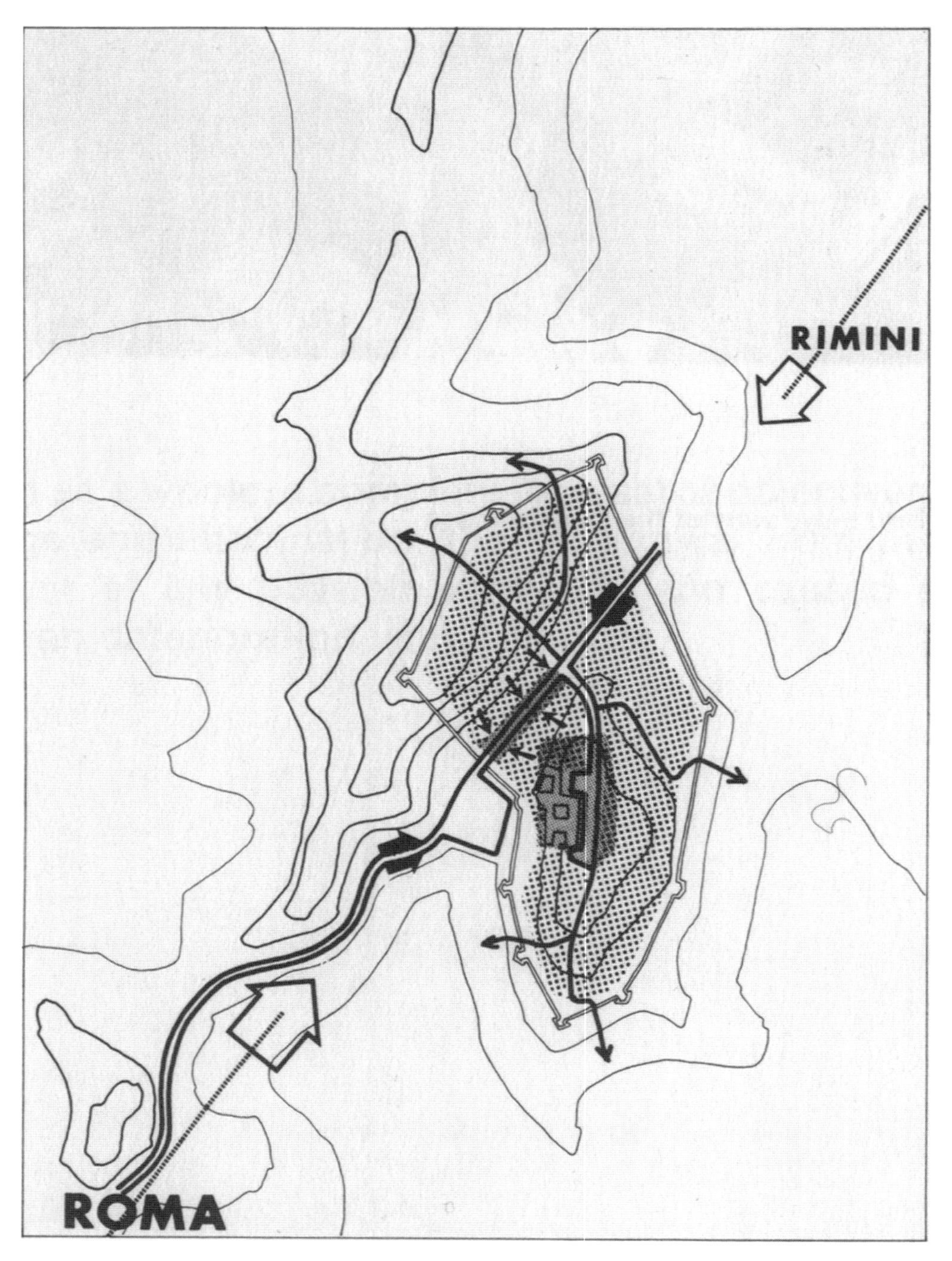

FIGURE 4.8b Urbino. 'The Renaissance town'. From De Carlo, *Urbino*, p. 98 (courtesy IUAV, Archivio Progetti, Fondo Giancarlo De Carlo)

Conclusion

In their plans for Assisi and Urbino, Giovanni Astengo and Giancarlo De Carlo pursued the twofold goal of implementing state-of-the-art rationalist planning techniques while preserving the memory of the places with which they were confronted. Their approaches to these historic cities were divergent, and partly linked to their different ideas about the roles of architecture and planning in the transformation of the built environment. Astengo's plan staged a conflict between heritage and modernity, insisting on the need to draw a clear distinction between the new Assisi and its historic townscape, while preserving the latter in its physical and stylistic integrity. De Carlo's plan seemed to conceive of urban heritage not only as a built testimonial of the city's past but also as a resource, with historic architecture almost posited as a generating force for future architectural interventions.

But the ways in which the two plans considered the historic townscape also presented some interesting commonalities. For all their interest in planning methodology and their willingness to set up a standard and scientifically grounded technique of urban analysis, Astengo's and De Carlo's approaches to the study of the historic city insisted above all on the role played by the planner's personal intuition in recognizing the fundamental traits of the built environment. Despite their many attempts to provide abstract representations for the complexity of urban life – for example by means of statistics or diagrams – the plans ended up incorporating a wildly heterogeneous set of materials, some of which were rich in pictorial and literary suggestions. Finally, despite the international scope of their work and the ambition of each to produce a document that could serve as a model for future planning experiences, the two Italian planners seemed to consider the study of the historic city mostly as a locally grounded experience, something that involved a direct confrontation with a city's individual character and with locally accumulated traditions, narratives and identities.

Notes

1. This research is partly based on the personal papers of the two architect-planners, which are stored at the Archivio Progetti of the IUAV University in Venice (henceforth IUAV, AP). Catalogues of both archives have been published recently: Marin, A. (ed.) (2000) *Fondo archivistico Giovanni Astengo. Inventario,* Venice: IUAV; Samassa, F. (ed.) (2004) *Giancarlo De Carlo. Inventario analitico dell'archivio,* Padua: Il Poligrafo. I wish to thank Bruno Dolcetta and Alessandra Marin for granting me access to Astengo's papers and guiding me through the first steps of the research, as well as the staff of the Archivio Progetti for their invaluable support.
2. Tafuri, M. (1989) *A History of Italian Architecture, 1944–85,* Cambridge, MA, and London: The MIT Press, Chapter 4; Bonfanti, E. and Porta, M. (1973) *Città, museo e architettura. Il Gruppo BBPR nella cultura architettonica italiana 1932–1970,* Florence: Vallecchi, 156–64; Cohen, J.-L. (1984) *La coupure entre architectes et intellectuels, ou les enseignements de l'italophilie,* Paris: École d'Architecture de Paris-Villemin.
3. The plan was extensively published in a monographic issue of *Urbanistica:* Astengo, G. (1958) 'Assisi: salvaguardia e rinascita', *Urbanistica,* 27, 24–25: 9–132.
4. De Carlo, G. (1966) *Urbino. La storia di una città e il piano della sua evoluzione urbanistica,* Padua: Marsilio; English translation by Schaeffer Guarda, L. (1970) *Urbino: The History of a City and Plans for its Development,* Cambridge, MA, and London: The MIT Press (all subsequent quotations are taken from this edition). De Carlo was a member of Team 10 and organized a meeting of the group in Urbino to coincide with the publication of the plan: Samassa, F. (2005) 'Urbino 1966. Team 10 in crisis: to move or to stay?', in Risselada, M. and van den Heuvel, D. (eds), *Team 10. In Search of a Utopia of the Present,* Rotterdam: NAi Publishers, 141–3. The plan for Urbino was also awarded an Abercrombie Prize from the Union Internationale des Architectes in 1967.
5. Astengo was the editor of *Urbanistica* from 1953 until 1976. Di Biagi, P. (1992) 'Giovanni Astengo. Un metodo per dare rigore scientifico e morale all'urbanistica', in Di Biagi, P. and Gabellini, P. (eds),

Urbanisti italiani. Piccinato, Marconi, Samonà, Quaroni, De Carlo, Astengo, Campos Venuti, Rome and Bari: Laterza, 395–467; Ciacci, L., Dolcetta, B. and Marin, A. (2009) *Giovanni Astengo. Urbanista militante,* Venice: Marsilio and IUAV University.

6. IUAV, AP, Astengo Papers, FAS 28 (correspondence on the plan for Assisi, 1955–60). Cf. also Brunelli Astengo, M. (ed.) (1959) *Le vicende del PRG di Assisi. Cronaca, articoli e documenti di due anni: 1958–59,* Venice: IUAV. Astengo did not cease working on the region in later years. He had already become involved with the drafting of a general plan for Gubbio (1956–60) and later acted as a consultant on the regional plan for Umbria (1960–2).
7. Perin, M. (2000) 'Giancarlo De Carlo. Un progetto guida per realizzare l'utopia', in Di Biagi and Gabellini, *Urbanisti Italiani,* 333–94; Bunčuga, F. (2000) *Conversazioni con Giancarlo De Carlo. Architettura e libertà,* Milan: Elèuthera; McKean, J. (2004) *Giancarlo De Carlo: Layered Places,* Stuttgart: Axel Menges, 52–113. De Carlo was granted honorary citizenship of Urbino on his 70th birthday, in 1989.
8. Rossi, L. (1987) *Giancarlo De Carlo: architetture,* Milan: Mondadori; Mioni, A. and Occhialini, E.C. (eds) (1995) *Giancarlo De Carlo: immagini e frammenti,* Milan: Electa; Fuligna, T. (2001) *Una giornata a Urbino con Giancarlo De Carlo visitando le sue architetture,* Urbino: Comune di Urbino, Assessorato alla Cultura.
9. IUAV, AP, De Carlo Papers, Atti 029 (PRG Urbino).
10. Filoni, M. (2004) 'Livio Sichirollo', *Rivista di storia della filosofia,* 2: 577–93; Azzarà, S.G. (2009) *Politica, progetto, piano. Livio Sichirollo e Giancarlo De Carlo a Urbino, 1963–1990,* Ancona: Cattedrale. The importance of Sichirollo's role is evident from his correspondence with De Carlo: IUAV, AP, De Carlo Papers, varie. Sichirollo's many writings on the plan include Sichirollo, L. (1963) *Filosofia e politica, ovvero del significato di una città: considerazioni di un amministratore,* Urbino: Istituto Statale d'Arte; Sichirollo, L. (ed.) (1964) *Il futuro dei centri storici e il PRG di Urbino, I. Primi documenti di cultura e vita amministrativa urbinate,* Urbino: Argalìa; Sichirollo, L. (ed.) (1967) *Il futuro dei centri storici e il PRG di Urbino, II. Vecchi e nuovi protocolli (1963–1967) con altri testi di cultura e vita amministrativa,* Urbino: Argalìa; Sichirollo, L. (1972) *Una realtà separata? Politica, urbanistica, partecipazione,* Florence: Vallecchi.
11. De Carlo, G. (ed.) (1966) *La pianificazione territoriale urbanistica nell'area milanese,* Padua: Marsilio.
12. Dolcetta, B. (1991) 'L'esperienza di Assisi', in Indovina, F. (ed.), *La ragione del piano. Giovanni Astengo e l'urbanistica italiana,* Milan: Franco Angeli, 103–19.
13. Astengo, 'Assisi', 11, 79–85.
14. De Carlo, *Urbino,* 119, 22–4.
15. His book *Questioni di architettura e urbanistica,* published in Urbino to coincide with the presentation of the plan, included a chapter against functional zoning: De Carlo, G. (1964) 'Fluidità delle interrelazioni urbane e rigidità dei piani di azzonamento', in De Carlo, G., *Questioni di architettura e urbanistica,* Urbino: Argalìa, 7–28.
16. Sichirollo, *Una realtà separata?,* 50.
17. De Carlo, *Urbino,* 34; Astengo, 'Assisi', 105.
18. Ciucci, G. (1982) 'Il dibattito sull'architettura e la città fascista', in Zeri, F. (ed.), *Storia dell'arte italiana. Il Novecento,* Turin: Einaudi, 263–391; Zucconi, G. (1989) *La città contesa. Dagli ingegneri sanitari agli urbanisti (1885–1942),* Milan: Jaca Book.
19. The latter quotation is taken from the script of *Città e terre dell'Umbria,* a short film directed by Astengo in 1961 for the exhibition held in Turin on the centenary of Italy's unification. On this occasion, Astengo also took part in the design of the section of the exhibition dedicated to the Umbrian cities: cf. (1961) *Italia 61. La celebrazione del primo centenario dell'Unità d'Italia,* Turin: Comitato nazionale per la celebrazione del primo centenario dell'Unità d'Italia, 428–31.
20. Astengo, 'Assisi', 46. Translated by Filippo De Pieri and Harriet Graham.
21. Ibid., 46, 49.
22. De Carlo, *Urbino,* 77.
23. Ibid., 87.
24. See, for example, Cristofani, A. (1866) *Delle storie d'Assisi, libri sei,* Assisi: Tip. Sensi; Rotondi, P. (1950–1) *Il palazzo ducale di Urbino,* Urbino: Istituto Statale d'Arte per il libro.
25. Astengo, 'Assisi', 79.
26. Cf., for instance, IUAV, AP, Astengo Papers, FOT 012, FOT 015, FOT 018.
27. Interestingly, at the same time Muratori's typological study of Venice analysed the less visible, if not virtually hidden, recurrent traits of the built environment: Muratori, S. (1959–60) *Studi per una operante storia urbana di Venezia,* Rome: Istituto Poligrafico dello Stato.

28. It is significant in this respect that the cover of the issue of *Urbanistica* dedicated to the plan featured a detail from an early Renaissance painting (Fra. Angelico's *Deposition from the Cross,* 1432–4).
29. The centrepiece of the critical analysis of the historic city was the 'Census of architectural and planning values' (Astengo, 'Assisi', map 2), a map that covered the whole built area of the historic centre and proposed a classification of each building according to a scale that ranged from 'Roman or medieval or other buildings of highest architectural value' to 'recent building of no architectural value and in striking contrast with the environment'. The map was intended as a guideline for future architectural interventions in the city centre.
30. Astengo, 'Assisi', 52–61. The use of photographic manipulation was not unusual for *Urbanistica* in the years of Astengo's editorship.
31. *Il volto di Assisi,* in Astengo, 'Assisi', 14–16.
32. The plan included a specific section on architectural restoration ('Risanamento e restauro', 108–16), where a similar set of photographs guided the reader to an understanding of Assisi's minute architectural details, distinguishing between 'Ancient medieval elements', 'Authentic medieval elements', '15th-century additions, 'Architectural elements from the 16th, 17th and 18th centuries', 'Authentic but dilapidated architectural elements that should be conserved in their *status quo*', 'Acceptable and feasible restorations' and 'Awkward restorations'.
33. De Carlo, *Urbino,* Chapter 5, 'The Structure and Form of the City', 74–97.
34. Castiglione, B. (Preti, G. ed.) (1965) *Il libro del Cortegiano,* Turin: Einaudi.
35. Mumford, E. (2000) *The CIAM Discourse on Urbanism, 1928–1960,* Cambridge, MA, and London: The MIT Press, 150–2, 201–15.
36. Rossi, A. (1966) *L'architettura della città,* Padua: Marsilio, English translation by Ghirardo, D. and Ockman, J. (1982) *The Architecture of the City*, Cambridge: MIT Press. Vasumi Roveri, E. (2010) *Aldo Rossi e L'architettura della città. Genesi e fortuna di un testo,* Turin: Allemandi.
37. De Carlo, G. (1968) 'Funzione, struttura e forma del centro storico', in *Atti del convegno sui centri storici delle Marche,* Rome: De Luca, 43–50; De Carlo, *Urbino,* 125–6.
38. Molinari, L. 'The spirits of architecture: Team 10 and the case of Urbino', in Risselada and van den Heuvel, *Team 10,* 301–6.
39. A friend of De Carlo since the 1940s, Albe Steiner (1913–74) was part of the group of intellectuals whom De Carlo periodically met in the holiday resort of Bocca di Magra. Longhi, G. (ed.) (2003) *Albe Steiner. La costruzione civile del progetto,* Rome: Officina; Steiner, A. and L. (2006) *Albe Steiner,* Milan: Corraini. Whether Steiner also contributed to the design of the diagrams is uncertain, but any temptation to assign him too strong a role would be contradicted by his own admission that he was involved in the preparation of the book only after May 1964, when the drawings were already finished. Steiner, A., intervention on the debate 'Il futuro dei centri storici' (November 1966), in Sichirollo, *Il futuro dei centri storici e il PRG di Urbino, II:* 231–3. De Carlo's original drawings for the diagrams can be found in IUAV, AP, De Carlo Papers, pro-014, 1.
40. De Carlo, *Urbino,* 58, 69, 70–4.
41. De Carlo, *Urbino,* 98–101.

5

'THE FIRST MODERN TOWNSCAPE'?

The Festival of Britain, townscape and the Picturesque[1]

Harriet Atkinson

Introduction

In August 1951 *The Architectural Review* proclaimed that 'the South Bank Exhibition may be regarded as the first modern townscape'. In a special Festival of Britain edition of the journal, its editors triumphantly declared that the layout of the Festival's South Bank Exhibition in London 'represents that realization in urban terms of the principles of the Picturesque in which the future of town planning as a visual art assuredly lies'.[2] They praised the way buildings outside the boundaries of the exhibition had been visually connected with those inside, the way the exhibition's buildings linked old with new. In doing so *The Architectural Review*'s editors claimed the Festival of Britain as a fulfilment of their long-running campaign for visual planning and a Picturesque revival. But was this indeed the intention of the Festival's architects, or simply self-justification on the part of the *Review*?

The Festival of Britain was a nationwide series of celebrations, held from May to September 1951. Clement Attlee's Labour government had agreed to mount the Festival after a committee report in 1946 recommended that such events would serve a useful function.[3] As well as giving the British people a lift after the war, they had the potential to boost international trade and tourism, to show Britain as a model democracy in an increasingly uncertain world at the start of the Cold War and to be a visible demonstration of the rapid reconstruction of the country after the destruction of war. As a date, the year 1951 was important symbolically, marking the centenary of Prince Albert's Great Exhibition, held in London in 1851, a moment when Britain was the workshop of the world and Britain's Empire extended in all directions. The 1851 Exhibition had put the production of 'all nations' on comparative display. By 1951 Britain's international influence was waning, with the Attlee government's granting of self-government to India, Pakistan, Burma and Ceylon from 1947 to 1949. Britain's finances had also been severely limited by war, and it was considered that an international trade fair would in fact be of limited use in promoting national manufacturing. The government therefore decided the Festival should focus on national achievements alone.

Festival of Britain planning started in earnest in 1947. Herbert Morrison was the minister in Attlee's government tasked with seeing the Festival through. He appointed Gerald Barry as Director-General. As ex-editor of the Liberal broadsheet *News Chronicle,* Barry was a shrewd choice. He believed strongly that architects and designers had a crucial role in reconstructing Britain by building it in three dimensions. As editor of the *News Chronicle* from 1936, Barry had considered it important that readers be given the chance to engage with contemporary architectural practice. The paper's open architecture competitions and architectural columns espoused radical solutions; the first of their kind, they caught the imagination of a public wanting to know more about how its country was being rebuilt.

While Morrison and Barry fought to justify the Festival in Parliament or to the press, the work of designing this exemplary Britain, to be displayed to the world in 1951, was to be led by architects and designers. Barry set up committees to give form to the Festival. The Exhibition Presentation Panel planned the exhibitions' design. The Panel included the 38-year-old Bartlett-trained architect Hugh Casson as Director of Architecture and several other architects and exhibition designers. The ambitions of demobbed, recently trained architects to build a brave new world after the war had largely been frustrated. The combination of materials shortages, building licences, the worsening international situation and the economic downturn meant only a few architects had had much actual building experience. Here in the Festival of Britain was their chance to demonstrate the pioneering role that architects and designers might have in rebuilding a post-war Britain. Many of the Festival's prominent designers were members of MARS (the Modern Architectural Research Group) and of its European counterpart CIAM (Congrès Internationaux d'Architecture Moderne). While advocating modernist architecture in Britain and Europe, these designers were also interested in Britain's architectural history.[4]

Eight government-sponsored Festival exhibitions were planned across the four nations of England, Scotland, Wales and Northern Ireland. These included a spectacular centrepiece at London's South Bank (Figure 5.1). This location, on the south bank of the River Thames, was chosen as the site for the Festival exhibition after months of deliberation. It was close to the historic centre of London and had already been marked out in 1943 and 1944 plans as ripe for development into a cultural quarter.[5] This was, in fact, returning the South Bank to a short-lived earlier use as the location of several pleasure gardens during the eighteenth century. With the site's rapid transformation from a wasteland populated by the old Lion Brewery (built in 1836–7), disused industrial buildings and slum dwellings into a 'magical city' populated with new buildings and landscapes, the South Bank was a vivid metaphor for reconstruction in action.[6] In Gordon Cullen's first airing of townscape ideas, his 'Townscape Casebook' published in *The Architectural Review* of December 1949, amongst a number of urban and rural examples the South Bank was used to illustrate concepts of 'Exposure' (the 'sense of release' felt when suddenly out in the open) and 'Ornament of Function' (the decoration of a scene with flags, colours and objects).[7]

Beyond London's South Bank there were to be Festival exhibitions in Belfast and Glasgow, at London's Battersea and at London's Science Museum in South Kensington; there were also two travelling shows and a 'Live Architecture' Exhibition at Lansbury in London's Poplar. Two thousand or so small events were planned up and down the country, ranging from town planning exhibitions in New Towns under construction to local arts festivals, new bus shelters and village fêtes.

FIGURE 5.1 The South Bank Exhibition at dusk: long view with Dome of Discovery and Skylon. Courtesy of University of Brighton Design Archives.

The Picturesque at the South Bank

The Festival's organizers decided the eight major exhibitions would be united by the theme 'the land and the people of Britain'. Linking the national topography with British national character and showing the dependence of British people on their land, this theme was to be elaborated within the exhibition displays, as well as in the treatment of the buildings and landscaping of exhibition sites (Figure 5.2). At the South Bank many visitors entered through the Station Gate. Once through the turnstiles, they stood in the exhibition site's largest open space, the concourse, facing the Thames straight ahead, with long views to the Dome and Skylon. The two-part structure of 'land' and 'people' allowed designers to make a virtue of the newly built Thames-side industrial site, divided halfway through by a railway embankment. Upstream – to the left – was the 'Land of Britain', overseen by exhibition designer

Misha Black; downstream – to the right – was the 'People of Britain', overseen by architect Hugh Casson. Guides and signposts bid visitors to follow the circulation route, to enter the upstream 'Land of Britain' by walking beneath a blue concave aluminium cone, the portico signalling the start of that section designed by architect H. T. ('Jim') Cadbury-Brown. The South Bank's circulation route invited visitors to follow a narrative through displays that started from the moment when the earth was formed, continuing through the moment when its elements were used to produce power and concluding when people used power to make objects used in everyday life.

From the start of the discussions, Festival architects revived eighteenth-century ideas of the Picturesque to shape the landscapes of these public exhibitions. Eye-pleasing arrangements of buildings in landscapes, experiments with visual variety and consciously scenic ensembles were created that had hitherto been associated with the design of the country estates of England. Speaking of the South Bank Exhibition's design in 1951, Casson said,'[O]n purpose it did not have the symmetry and the repetitive grandeur of some other great cities and their exhibitions. It was planned intimately, like rooms opening one out of another'.[8] Festival designer Gordon Bowyer confirmed Picturesque considerations were at the forefront of Festival designers' minds, recalling Casson walking around the South Bank site with landscape designer Peter Shepheard and expressing delight when planning alignments had been lost during the building process, producing a pleasing irregularity.[9]

The impact of the Picturesque at London's South Bank site was partially achieved through a kinetic relationship between displays and people: the action of visitors moving through the

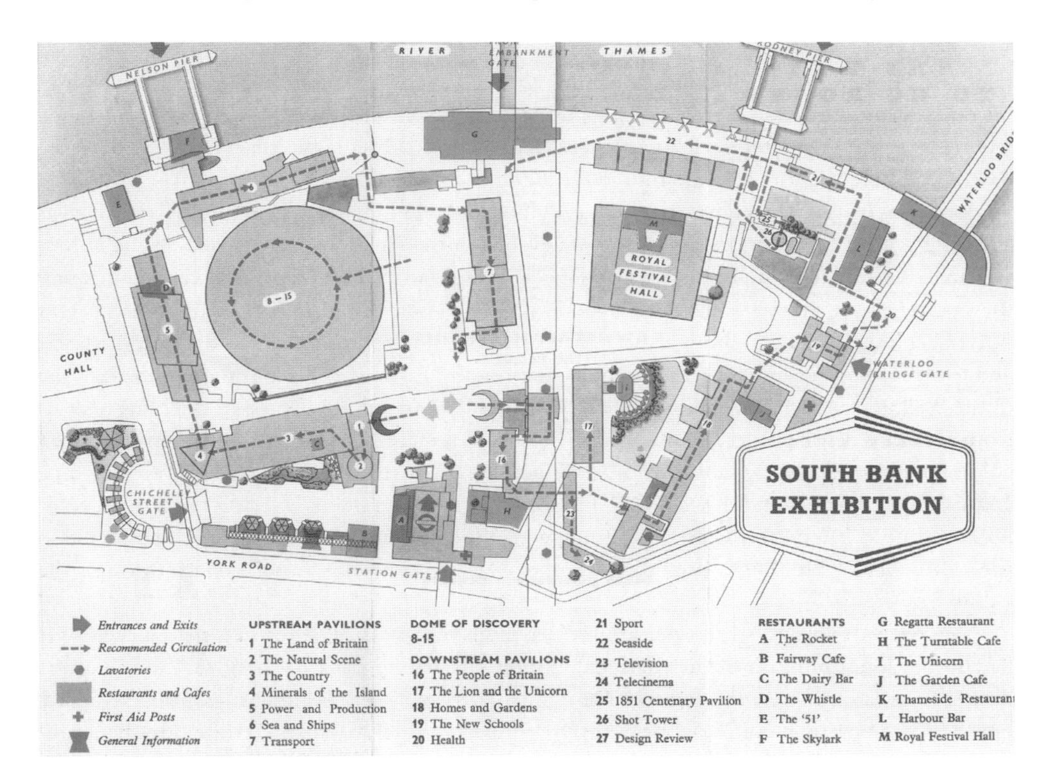

FIGURE 5.2 South Bank map, from South Bank Exhibition Guide. (London: HMSO, 1951.)

exhibition spaces, from dark and dim to brightly or naturally lit; from enclosed, narrow corridors to colossal, open inside and outside spaces. At the South Bank buildings were set over many levels, with steps and walkways taking visitors up and down, to gaze over constantly changing views. There was much incident: fountains played and ducks bathed, helium balloons were let off into the sky above with tickets attached for lucky passersby, people stopped to catch their breath while watching cascading water sculptures. Some ate Cornish pasties while sitting on chairs set about the site, relaxed in open-air cafés in the shade of a coloured umbrella or sat gazing out across the Thames at the magnificent views towards the Houses of Parliament, Waterloo Bridge and St Paul's. The South Bank was crowded in places with densely planted shrubs and trees, dotted with concrete planters holding jauntily coloured flowers. Even the Dome of Discovery, this icon of British modernism, this harbinger of the future, operated within the idiom of the Picturesque. It appeared at first to be a perfectly symmetrical disc but changed in view and proportions as visitors walked around it, with steps allowing for changes of level and vantage point.[10]

The South Bank's Picturesque worked both by day and by night. At dusk, the Festival site was designed to undergo a transformation using a combination of fountains, dancing, music and lighting to enable visitors to continue at the site after dusk. Floodlighting of Festival buildings and structures glimpsed across the Thames was designed to create a magical environment. Lights set in the paving slabs of the site illuminated pathways so that visitors could walk, sit looking out at the view or dance to the strains of music penned for the Festival. Fairy lights in red and white set in trees were switched on, as were lights in the so-called Fountain Lakes, designed by Jim Cadbury-Brown, who had been inspired by a visit to see the fountains at Versailles, and supported by Gerald Barry, who had visited Rome in 1950 to look at the fountains and floodlighting. Fire-making technology and gas flares were incorporated into them to create an additional mist. Artificial lily-pads with lights set into them, dotted about in ponds, were also turned on after dark to help visitors navigate through the space and to allow the daytime Picturesque to continue into the evening.

As well as bright floodlighting, which provided a vivid night-time structure for the Festival, allowing visitors to see the extent of the site even after the pavilions were closed, strong colour was an extraordinarily important part of the Festival spectacle. This colour was carried through the open-air sites, as well as into displays and objects. The exhibition sites 'blazed with bright nursery colours' – carried though the signage, the blue and red umbrellas at the outdoor cafés and eye-catching screens such as the molecular one by Edward Mills and the tetrahedral one of the Architects Co-Partnership.[11] These performed the practical purpose of shielding visitors from the grey desolation of the industrial wasteland remaining beyond the site. In this sense, colour performed a structural purpose, compensating for missing buildings and unwelcome empty spaces. But beyond this, colour had a strong symbolic value, signifying modernity, vitality and simple 'gaiety'. As Casson commented, '[T]he colours were as carefully considered as the forms of the buildings', providing another way of achieving visual variety through the exhibition sites.[12]

The Festival's layout, structured around pavilions and with features like the Fairway, was based on a long tradition of international exhibitions, dating back to the great Paris Expositions of the late nineteenth century. The South Bank site's Picturesque effect was achieved, however, by departing from axial Beaux Arts symmetry, which had been favoured in the layout of international exhibitions such as the 1867 and 1889 Paris Expositions. Casson and his team instead chose a design based on routes that meandered around the site, strongly influenced by three recent exhibitions that had experimented with ideas of identity of place in newly built buildings and landscapes: Asplund's landscaping of the 1930 Stockholm Exhibition, held at Djurgården; Hans Hoffmann's landscaping of the 1939 National Exhibition (the 'Landi'), held on two banks of the River Limmat in Zurich; and the 1948 Exhibition of the Lands Regained, held in the Polish city of Wrocław, all known to the Festival architects through their coverage in architectural journals (Figure 5.3).

The design of the South Bank landscape was planned as part of the wider architectural scheme. H. F. ('Frank') Clark was the lead landscape designer, assisted by Maria Shephard. Peter Shepheard, Peter Youngman and Jim Cadbury-Brown worked with them (Figure 5.4). Instructions to Festival landscape designers dictated that '[t]he use of colour and plant forms should be in the spirit, though not necessarily in the manner of the 18th century landscape garden, which was designed to evoke emotion, and awaken dreams'.[13] Planting schemes were of special importance at the South Bank in order that the site would become not only a landscape but a green one, fulfilling Barry's idea that it would really be 'alive' for the summer of 1951 so that visitors might experience a sense of being uplifted as on a country walk, even while in the heart of the city. Planting across the South Bank took many different forms. Some areas were made to appear like the undergrowth around wild rivers or streams. Youngman designed a Moat Garden, a practical landscaping device aimed at controlling crowd flow, which appeared like a riverbed with large stones, shale and water, surrounded by bushes and plants. Snaking around one of the café areas, this stretch of water drew the eye into the immediate environment and away from the buildings just outside the site. On other parts of the South Bank, planting was more manicured. Around the Homes and Gardens building, lawns, wall-mounted stoneware planters and a sunken garden by Shepheard dotted with kitchen herbs gave the impression of a lovingly tended domestic garden. Meanwhile, Cullen was commissioned to design a modest pocket garden with rubble stone walling on a terrace outside the South Bank's Homes and Gardens.

The replacement of trees after their loss during wartime deforestation, to make way for roads and aerodromes, was a symbolic act of reconstruction. Landscape architect Peter Shepheard described his conviction that trees could be used in a process of reconstruction: 'There is hardly a town anywhere which has not some scar of industry or railway yard, gasworks or speculative building, which could be healed by the careful planting of the right trees in the right places'.[14] Using trees to 'heal' scars was linked with a deep, quasi-spiritual belief in the capacity of elements from nature to cure the urban condition. This idea was followed through with great ambition. Semi-mature, 30-foot-high trees were transplanted to the South Bank in the autumn of 1950. A thicket of rhododendrons from a Rothschild estate on the edge of the New Forest was lifted and brought to the site, providing a sheltered area without the need for a wall. Across Britain, tree planting was also a significant part of the Festival, drawing together villagers and townspeople. Underlying this activity was an idea of locating beauty and particularity of place, even in places that had been demolished and entirely rebuilt. Even where a tabula rasa existed as a result of total destruction, or there was deep industrial scarring, places could notionally be returned to their 'proper' form, an original form that identified the *genius loci,* a kind of prelapsarian state. Festival landscape architect Shepheard had a wider belief that the land had an intangible but specific 'character'. 'The landscape architect is concerned with the existing site; with its character and the *genius loci*. Buildings also, and especially modern buildings, must, of course, fit the site and acknowledge its character; but landscape is the site', he explained.[15] Festival planting, which brought greenery and countryside features to the

FIGURE 5.3 Exhibitions celebrating land and people: (Above) Stockholm Exhibition, 1930. Asplund's landscaping was particularly praised by the British architectural press. (Below, left) Cover of the catalogue of the Landi, Zurich, 1939, showing the exhibition's site on the banks of the River Limmat. (Below, right) Tower of zinc buckets at the Exhibition of the Lands Regained, Wrocław, 1948.

South Bank, helped visitors link this urban exhibition site with the world beyond. It also provided a pattern book showing how the country might be integrated with the city. 'The main beauty of [the South Bank Exhibition] was to show people that an urban site is just as much fun as a rural site', Shepheard later commented. He went on,

> Built into the English mind is the idea that peace is in the country and, of course, much funnier ideas that 'God made the country and man made the town', which of course is horribly untrue. We wanted to demonstrate that.[16]

Architect Jim Cadbury-Brown was responsible for the design of the Origins of the Land pavilion and the landscaped area around it. He was working at the same time on hard landscapes for Harlow New Town.[17] His work at the South Bank allowed him to extend the internal display about geology seamlessly onto the exterior landscaping. These displays were not intended to look entirely naturalistic: Yorkshire stone sat beside square-formed Forest of Dean stone and red sandstone, a geological impossibility. Instead, they made strong references to the British land beyond that visitors could easily pick up on. They employed the materials that they represented – in this case a stony outcrop in the British landscape – to tell their own story. Cadbury-Brown recalled visitors on a day-trip to the South Bank becoming so convinced by the earth-strewn boulders clustered around the mouth of his pavilion that some started picnicking on them, something that was firmly clamped down on by attendants fearing the display would be ruined.

In the Festival's modernist buildings, architects also used stone brought from other parts of Britain in the buildings, landscaping and paving, alongside concrete, brick and new materials. The Royal Festival Hall, designed by Robert Matthew and Leslie Martin with Peter Moro of the London County Council's Architects' Office, made much use of geological samples. The Hall's large external columns were faced with a Derbyshire marble known as 'Derbydene', a polished variety of carboniferous limestone composed mainly of fossil debris. The columns exposed 'occasional corals and brachiopods embedded in a matrix of comminuted fragments and lime mud', carefully left in to hint at its origins. This direct link with the land was even more evident when the quarry source of this 'prehistoric slab, with interest close-to' ran out and a section at the top of the building could not be covered.[18] This detail was picked up on the front cover of *The Architectural Review* for June 1951, which showed a close-up of the fossilized limestone used at the Royal Festival Hall with a caption: 'The Cover shows the genius loci of the Royal Festival Hall as it manifested itself to Gordon Cullen while he was examining the Derbyshire marble walls of the foyer'.[19]

The South Bank's walkways used a wide variety of igneous, metamorphic and sedimentary rocks displaying textures, structures and fossils hinting towards their origin and evolution. Meanwhile, visitors walked through a succession of pavilions devoted to the theme of 'the land', all by different architects: The Land of Britain, The Natural Scene, The Country and The Minerals of the Island. While being told about the land in the exhibitions' content, visitors to the site were invited to enact their relationship with the land through their exposure to the site itself.

The potential of the Picturesque at the South Bank was not seen only as carried within buildings and landscapes (Figures 5.5, 5.6). The lettering and typography at the Festival's South Bank were considered crucial to the ensemble. Before the Festival, designers such as Milner Gray had participated in planning research in the early 1940s into the way that street lettering could be included within wider planning schemes.[20] Led by planner W. G. Holford and Sir Stephen Tallents, the research focused on creating standards for legibility, materials and siting of signage and on respecting the architectural and civic character of a locality by allowing for

FIGURE 5.4 South Bank Exhibition landscape: (Above, far left) The Country with farmer's wind-pump as a rotating model of sun and moon. (Above, middle) Concrete flower pots and Ernest Race chairs in front of the Dome. (Above, far right) The Unicorn café with planters. (Middle, left) Lawns outside Homes and Gardens. (Middle, right) Herbs set in paving outside Homes and Gardens. (Bottom, left) Pool and Jacob Epstein's *Youth Advancing* outside Bronek Katz and Reginald Vaughan's Homes and Gardens. (Bottom, right) Wall-mounted pots and planters outside Homes and Gardens designed by Maria Shephard. Courtesy of University of Brighton Design Archives.

variation, within reason. In the same vein, Cullen had demonstrated where painted advertisements fitted well within their environment in a 1949 study of publicity within the townscape for *The Architectural Review*.[21] In pursuit of a unified look for all of the Festival's signing and text – in particular the 'titling of buildings' – the Typography Panel was set up to advise on this aspect of the exhibitions. At the South Bank, lettering was chosen that lay flat against buildings and did not project in three dimensions from the surface to which it was applied – although it was designed as if in three dimensions. This was considered universal enough to fit every Festival context. The decorative Victorian styling of the text contrasted with the minimal decoration of the signage, litterbins and buildings to which it was applied, an effect that the alphabet's designers had specifically hoped to achieve.[22] Nikolaus Pevsner praised the lettering schemes for the way they contrasted with the buildings around them. The distinct difference between the florid, Victorian, 'over-robust' Festival lettering and the experimental modernism of the buildings with 'transparency of glass, slim and vigorous steel struts, and thin aluminium lattice work' as background to the lettering was, he thought, integral to their design. The contrast was, for Pevsner, its ingenuity and an important and central aspect of the Picturesque presentation and the visual variety of the Festival that he had argued for elsewhere.[23] Aside from lettering, Pevsner also saw the Picturesque in the way paintings and sculptures had been laid out at the South Bank, as well as in 'minor elements of external furnishings'.[24]

FIGURE 5.5 Festival lettering, 'A Specimen of Display Letters Designed for the Festival of Britain 1951', designed for the Typography Panel of the Festival of Britain in 1951 for distribution to architects and designers showing lettering within the Festival townscape, p. 29. Courtesy of University of Brighton Design Archives.

The Festival and *The Architectural Review*

The Architectural Review's August 1951 edition was devoted to enthusiastic coverage of the Festival of Britain at London's South Bank.[25] A foreword announced that the South Bank Exhibition had 'triumphantly demonstrated the vitality of contemporary British architecture and should have a worldwide influence'.[26] Anonymous articles addressed 'The Exhibition as Landscape' and 'The Exhibition as a Town Builder's Pattern Book', while editor J. M. Richards wrote about 'The Exhibition Buildings' and assistant editor Gordon Cullen's 'South Bank Translated' described how future planners might use principles from the exhibition in the permanent development of the area. But how much was the *Review* in tune with what the Festival's designers were intending to achieve?

Since the 1930s, *The Architectural Review* had become the locus of vigorous debates about new architectural movements. The journal, which was read by architects and designers alike, was extraordinarily influential on the generation who designed the Festival. H. F. Clark, landscape consultant for the South Bank Exhibition, was one Festival designer with close links to the *Review.* He had written an article in the *Review* entitled 'Lord Burlington's Bijou, or Sharawaggi at Chiswick' in 1944, publicly endorsing the journal's agenda for the Picturesque.[27]

Hugh Casson, the Festival's Director of Architecture, also shared much common ground with the *Review.* He had worked as an architectural journalist from 1937, starting out on *Night and Day* and writing on subjects that strongly echoed themes favoured by the *Review*'s editors. Casson's seminal 1945 book *Bombed Churches as War Memorials,* printed by *The Architectural Review*'s publisher, Architectural Press, had appealed for the preservation of the ruins of some churches whose rebuilding in the centres of congested cities could not be justified but whose disappearance would sever people's link with their past. It was published around the time when the *Review*'s editor J. M. Richards and the artist John Piper were both also writing about the treatment of ruins in the landscape.[28] Before Festival planning started, Casson had already shown strong sympathies with the *Review*'s calls for a Picturesque revival, speaking up in support at a talk at the Architectural Association in November 1945 at which Pevsner had developed his theory of revived Picturesque.[29] Casson was a close friend and admirer of H. de C. Hastings, the main driving force behind *The Architectural Review*'s Picturesque campaign from the 1930s onwards.[30] In a later letter to Hastings, Casson declared, 'I regard myself your creation'. Casson later recalled taking Hastings around the South Bank site and being delighted when he commented that 'it was something like I had always wanted'. He declared, '[Hastings] was my guru and certainly the guru of the South Bank Exhibition'.[31] Casson was perhaps more enamoured with the Picturesque than his Festival colleagues were, however. Gerald Barry, the Festival's Director-General, admirer of orthodox architectural modernism, commented in 1949, 'The vogue for sharrawaggy [*sic*] – the natural reaction from a too-stark functionalism – can be overdone', directly referring to the campaign for Picturesque planning that had been waged in the pages of *The Architectural Review* through articles such as 'Exterior Furnishing or Sharawaggi' of 1944.[32]

The *Review*'s editors and the Festival's design team were also linked through their mutual membership of MARS. A number of Festival designers – Wells Coates, Jim Cadbury-Brown and Frederick Gibberd, for example – were members of MARS. Many MARS members were also part of CIAM, which had been formed in 1928 by, among others, Le Corbusier and Sigfried Giedion. During a period when there was limited scope for international travel, the British element was particularly influential through their hosting of CIAM meetings in 1947 and 1951. From 1947, *Architectural Review* editor Richards led MARS and shifted the group's direction away from its pre-war functionalist concerns towards his own philosophical interests in the aesthetic appeal of modern architecture for what he described as the 'Common Man'.[33] From 1948 to 1951 Richards was also a member of the Festival's Council for Architecture, Town Planning and Building Research, which oversaw the development of the Festival's 'Live Architecture' Exhibition at Poplar's Lansbury Estate.

The Picturesque at the Festival's exhibition of live architecture at Lansbury

The Architectural Review's discussion of the Picturesque and townscape at the Festival was largely focused on the South Bank, but the design of the new Lansbury Estate also operated within this visual idiom. The Lansbury Estate, the Festival's 'Live Architecture' Exhibition, was coordinated by the London County Council (LCC) and built in a blitzed area of east London. Lansbury was planned as one of eleven 'neighbourhood units'. The concept of 'neighbourhood units' had been put forward by Patrick Abercrombie as a model for structuring tight communities within larger planning masses. While the idea of the 'neighbourhood unit' had originated within North American planning debates, in the British context they were essentially imagined as linked villages.[34] Loosely defined as a small cluster of houses with the central focus of a church and an area of common land or pond, villages had repeatedly been cited as a model of a virtuous community for British planning. Post-war housing minister Aneurin Bevan had likened housing estates to 'modern villages', where all classes would live in harmony.[35] Meanwhile, planners such as Thomas Sharp in *The Anatomy of the Village* had sought to encapsulate the spirit and atmosphere of the village in order that it could be transferred to new, urban developments.[36]

The planning of the Festival's Lansbury Estate, built on a site that had been completely devastated during the Second World War Blitz as a result of its proximity to London's docks, reflected a belief in the existence of a continuing character or *genius loci* embedded in place, regardless of its current appearance. This influenced the LCC's instruction to Lansbury's various contributing architects to focus on using the slate, brick and stone that were characteristic of that part of east London. In addition, the way new housing was laid out at Lansbury reflected this preoccupation with manufacturing a sense of local particularity. At Pekin Close, designed by architects Bridgwater and Shepheard, houses were set in two-storey terraces made from brick with tiled, pitched roofs, each with gardens, set in a pedestrianized cul-de-sac, cut off from traffic by bollards. These houses would later sit in the shadow of the newly built Roman Catholic Church of St Mary & St Joseph (designed by Adrian Gilbert Scott, built 1950–9). Social housing at Lansbury mimicked the atmosphere of an intimate village by setting housing in leafy areas, predominantly low-rise and small-scale, many with their own gardens or else immediate access to green space. Access between clusters of houses was through a succession of green, landscaped spaces that were closely integrated and acted like village greens.

Through a rather extraordinary imaginative leap, the public authorities mounting the Festival at Lansbury saw themselves as the inheritors of the private landowners who had formulated the Picturesque two centuries earlier. Speaking at CIAM 8 in 1951 about their work at Lansbury, LCC coordinating planner Arthur Ling asserted that they saw themselves as taking on the mantle of eighteenth-century private landowners and thereby becoming twentieth-century patrons. According to Ling, while not wholly supporting the architectural methods of their forebears, they nevertheless owed a debt for the prevailing system of 'large-scale ownership

FIGURE 5.6 Festival lettering, 'A Specimen of Display Letters Designed for the Festival of Britain, 1951', p. 8. Courtesy of University of Brighton Design Archives.

of land', which 'allowed architects and planners to produce works of unquestioned merit'.[37] This imagined link between the LCC as a major landowner, working within a new system of democratic accountability, and an erstwhile land-owning generation, was at the heart of this public Picturesque, with its projection of historical continuity into the new British welfare state system. Indeed, Ling's words echoed Sharp's earlier comments in his 1946 *Anatomy of the Village* about public authorities taking over village planning from large landowners.[38]

The emphasis on creating environments for fully functioning communities in newly rebuilt areas was also manifest in the idea of 'mixed development', which saw new estates being populated with a cross section of age groups and a variety of groupings including families, couples and individuals. Festival architect Frederick Gibberd was a strong exponent of such developments, believing in the need for a social cross section and, of particular significance in this context, that mixed buildings produced an appropriate visual impact. 'Buildings with quite different formal qualities such as blocks of flats, maisonettes and bungalows are needed to provide "contrast" and "variety" in the "composition" of an area', he wrote.[39] The impact of this preference can be seen in Gibberd's schemes at Somerford Estate in Hackney (1947), at Harlow (1951) and in the Festival's 'Live Architecture Exhibition' at the Lansbury Estate.

This predilection for low-rise buildings on public housing estates that was shown at Lansbury, operating within a public Picturesque, also characterized the schemes singled out as having special merit in the Festival of Britain's 1951 Architecture Awards. The Awards set out to stimulate the 'creation of beauty', which its organizers saw as 'an appropriate form for celebrating the Festival throughout Great Britain'.[40] Award-winning schemes included Jury's Old People's Housing in Glasgow (designed by Glasgow City Council, 1949–50), where single-storey dwellings were set among mature trees with steeply pitched, tiled roofs, and terraced housing in Norfolk by Tayler and Green, with low roofs that fitted into the long, low lines of the local landscape. Low-rise buildings set in green space were also the favoured form for housing at Harlow New Town. At Harlow, in its earliest stages, Gibberd's only high-rise building was The Lawn (1951), a nine-storey point block. Standing, as it did, in isolation, The Lawn was more akin to a viewing tower that allowed a view down onto the Harlow Estate for those inside it, and a visual feature – like the 'eye stoppers' or focal points used as landscaping devices – for those below.[41] This newly completed building formed the centrepiece of Harlow New Town's Festival exhibition. Visitors were invited to climb to the ninth floor to view models, maps and drawings and then to step out onto the penthouse balcony to compare models with work-in-progress on-site (Figure 5.7). In the same spirit, Gibberd inserted a clock and viewing tower into his Festival designs for Lansbury's Market Place. He explained that the clock tower, which rose above the otherwise low-rise buildings of the market square, 'closes the long vista down the road leading to the square, and provides a contrast to the comparatively low shop buildings' and simultaneously closed the view to the desolate and unsightly stretch beyond.[42] In other New Town schemes-in-progress exhibited as part of the Festival of Britain, for example at Stevenage and Crawley, low-rise, pitched roofed terraces set carefully in landscapes were also the norm.

While *The Architectural Review* and, to a lesser extent, its sister paper, *The Architects' Journal,* were experimenting with the potential for a twentieth-century Picturesque, the revival of these ideas was certainly not met with universal approval. *Architectural Design* developed an antipathetic agenda, with writers such as E. A. Gutkind expressing their disappointment at this aspect of the Festival. Describing the use of vistas at the South Bank as 'empty ostentation' and tracing this back to the tradition of exhibition design pioneered in Paris by Haussmann, he dismissed the site as belonging to another age.[43] For several years after the Festival closed, the debate over whether there was any role for the Picturesque in the twentieth century was still raging. Amongst the antis was art historian Basil Taylor, who gave a series of broadcasts on the Picturesque in 1953. He criticized those in positions of authority in government arts, architecture and design agencies for their uncritical admiration for the Picturesque, mockingly suggesting that Herefordshire squires Richard Payne Knight and Uvedale Price, early champions of the Picturesque, would have been quite at home in this post-war world.[44] Despite this, the Picturesque revival was promoted as a necessary element of British reconstruction by commentators for several years, most prominently by Nikolaus Pevsner, who made this the subject of one of his 1955 Reith Lectures.[45]

The Lansbury Estate, at least in spirit, was a fulfilment of the 'Twentieth Century Picturesque' promoted by *The Architectural Review.* In appearance, however, it fell short of the mark, being criticized by ex-planning committee member J. M. Richards as 'completely orthodox in construction' and having a 'dull, less characterful external appearance' that 'followed in a nineteenth century building tradition', thus lacking the required innovation.[46] Gibberd himself, writing later, suggested he had not been happy with Lansbury, being 'too modest and lacking in exciting "architectural statements"'. This may, he thought, have been because its architects had been particularly keen to prioritize community needs above all. He concluded, however, that 'it was immensely important, and I suppose everyone who was in any way involved in the rebuilding of Britain was influenced by it'.[47]

Shopping Centre and Market Place seen from Chrisp Street

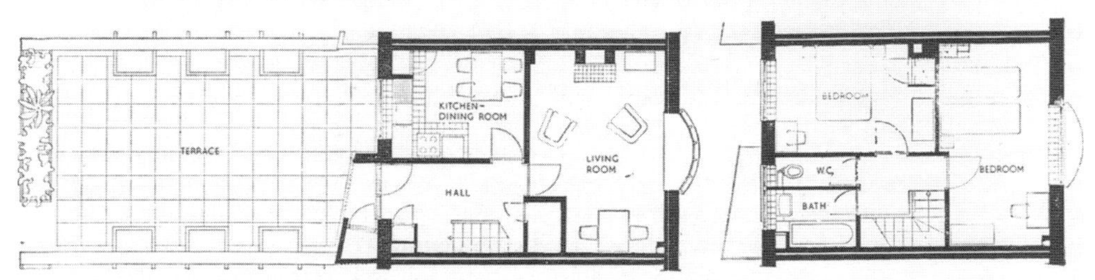

Plan views of a maisonette over a shop, showing lower and upper floors

FIGURE 5.7 Live architecture exhibition, Lansbury Estate, poplar: (Above) model of shopping centre and market place, with clock tower in foreground, seen from Chrisp Street. (Below) Model showing double arcade of shops and paved Market Place. Architect: Frederick Gibberd, from 1951 Exhibition of Architecture, *Poplar Guide* (London: HMSO, 1951).

The Festival of Britain closed on the last day of September 1951. On Monday, 1 October, only a few hours after it closed, the work of packing up and dismantling began. While the Lansbury Estate had been built to be permanent, at the South Bank Exhibition the only building intended to last was the LCC's Royal Festival Hall. By May 1952, with many of the Festival's buildings demolished, photographs began to appear in the press of a new South Bank Riverside Walk under construction in front of the Royal Festival Hall. This was now a vast open space. What would a post-Festival South Bank look like? One building was already in the pipeline. In the Festival year, the foundation stone was laid for a new National Theatre further downstream on the South Bank, on a site controlled by the LCC. Building planning did not start, however, until several years later, with the appointment of Denys Lasdun as architect, and the Theatre was not completed until the mid-1970s. The Queen Elizabeth Hall, a second concert hall designed by Higgs & Hill, was built near the Royal Festival Hall in 1967, and the Hayward Gallery, designed by Hubert Bennett, was completed in 1968.

Critics characterized these concrete buildings as 'Brutalist', using as a pejorative the phrase coined as a form of self-definition by the young architecture critic Reyner Banham in his article 'The New Brutalism' in *The Architectural Review* of 1955. The New Brutalists characterized themselves as a new generation and vociferously rejected the work of their forebears, who had been hung up on promoting historic styles. This was a manifesto for the future: the New Brutalism, they claimed, was an architecture that rejected historicism, an 'architecture of our time'.[48] Within this projection of the present and future, the Festival's experiments with the Picturesque became the counterpoint. In his essay 'Revenge of the Picturesque', written in 1968, the year the Hayward Gallery was completed, Banham criticized the Festival with its revived Picturesque as an escape from the real problems and opportunities of urbanization, a symptom of a xenophobia from which he and his contemporaries were at pains to distance themselves and utterly irrelevant to his contemporaries, relating more to the aesthetics of an older generation.[49] The younger generation whom Banham had seen as let down and ostracized by the dogma of the Picturesque revival included his friends, the architects Peter and Alison Smithson. In a 1976 essay, Banham again asserted his anti-Festival position, characterizing it as 'an overwhelming demonstration of the superiority of the English Picturesque tradition over all other planning dogmas'.[50] Within this debate, the buildings neighbouring the Royal Festival Hall were seen as a strong reaction against the Festival's tempered, peculiarly British modernism. Now, 60 years after the Festival, the South Bank's contextualized temporary townscapes no longer look a world away from what stands today. The Queen Elizabeth Hall's terraces offer multiple views towards the Thames and beyond, while the National Theatre with its carefully balanced horizontals and verticals provides a terraced urban landscape on the banks of the River Thames.

Notes

1. For a further elaboration of these ideas see Harriet Atkinson (2012) *The Festival of Britain: A Land and Its People,* London, New York: I.B. Tauris.
2. Caption for 'The Cover' of the *Architectural Review*, 110, 'Special issue on the South Bank Exhibition' (August 1951), p. 71.
3. Ramsden Committee investigations into holding a Festival, 1945–6, papers held at the National Archives, London.
4. Hugh Casson had, for example, written *Victorian Architecture* (1948, London: Art & Technics), exploring further the building of a period that many of his contemporaries had vilified.
5. Abercrombie and Forshaw's *County of London Plan* (1943, London: Macmillan) and Abercrombie's *Greater London Plan 1944* (1945, London: HMSO).

6. Phrase used by Misha Black (1961) 'What contribution did the Festival make to exhibition and display techniques?' *Design,* 149 (May), p. 51.
7. *The Architectural Review*, 109 (December 1949): 363–74. Cullen evolved these examples further in 'South Bank Translated' in *The Architectural Review* of August 1951, and they were published in his book *Townscape* (1961, London: Architectural Press).
8. In a valedictory Festival film, *Brief City* (1951), directed by Jacques B. Brunius and Maurice Harvey and narrated by Hugh Casson and Patrick O'Donovan.
9. Gordon Bowyer, interview with the author, 2004.
10. See Brunius and Harvey, *Brief City.*
11. Ibid.
12. Ibid.
13. National Archives, INF 12/255.
14. Notes about tree planting, National Archives, INF 12/255.
15. Shepheard, P. (1953) *Modern Gardens,* London: Architectural Press, p. 17.
16. Shepheard, P. (1997) *Architects' Lives, National Sound Archive* (14 of 18) 1997.07.02 to 1997.08.12.
17. As Cadbury-Brown explained, interview with author, April 2003.
18. Moro, P. (1995) *Architects' Lives, National Sound Archive* (8 of 15) 1996.8.01, 1996.8.14 and 1996.9.05.
19. *The Architectural Review,* 109 (June 1951).
20. 'Notes on proposed research into improvements in the design, manufacture and siting of street lettering', 26 May 1944, Gerald Barry archive (private).
21. Cullen, G. (1949) 'Townscape', section on publicity, *The Architectural Review,* 106 (December), p. 372.
22. Reilly, P. (1952) 'The Printed Publicity of the Festival of Britain', *The Penrose Annual: A Review of the Graphic Arts,* 46, p. 23.
23. Pevsner, N. (1952) 'Lettering and the Festival on the South Bank', *The Penrose Annual: A Review of the Graphic Arts,* 46: 28–30.
24. Pevsner, N. (2010) 'Visual planning and the Picturesque' in Aitchison, M. (ed.) *Visual Planning and the Picturesque,* Los Angeles, Calif.: J. Paul Getty Trust, p. 198.
25. *The Architectural Review,* 110, 'Special South Bank Edition' (August 1951). *The Architectural Review*'s editors were listed as J. M. Richards, Nikolaus Pevsner, Ian McCallum and H. de C. Hastings, with Gordon Cullen credited as assistant editor responsible for research.
26. Summary of Foreword, ibid., p. 71.
27. *The Architectural Review*, 95 (May 1944): 125–9.
28. Casson would join *The Architectural Review*'s editorial board in 1954.
29. As noted by Erdem Erten in 2004, 'Shaping the Second Half Century', PhD Thesis, Cambridge, Mass., MIT, pp. 58–9.
30. Hastings had taken on Pevsner as co-editor of the journal in 1942.
31. Hugh Casson, letter to H. de C. Hastings, n.d.; and Hugh Casson, interview with S. Lasdun, 1984; both quoted in Lasdun, S. (1996) 'H. de C. reviewed', *The Architectural Review* 200 (September): 68-72. J. M. Richards also acknowledged the Festival of Britain as a realization of Hastings's ideas: 'It was the physical embodiment of townscape policy', according to Susan Lasdun's 1975 interview with Richards.
32. Barry quoted in 'One Man's Day' (1949) *The Leader,* 9 July. 'The Editor' (1944) *The Architectural Review* 95 (January): 2–3.
33. Mumford, E. (2002) *The CIAM Discourse on Urbanism 1928–1960,* Cambridge, Mass.: MIT Press, p. 177.
34. See Pendlebury, J. (2009) 'The Urbanism of Thomas Sharp', *Planning Perspectives,* 24 (1): 3–27.
35. Ravetz, A. (2001) *Council Housing and Culture,* London: Routledge, p. 96.
36. Sharp, T. (1946) *The Anatomy of the Village,* London: Penguin. See also Stewart, C. (1948) *The Village Surveyed,* London: Edward Arnold.
37. Ling spoke about 'Satisfying Human Needs at the Core' at CIAM 8 in 1951. Recorded in J. Tyrwhitt, J. L. Sert, E. N. Rogers (eds) (1953) *CIAM 8: The Heart of the City,* London: Lund Humphries, p. 96.
38. Sharp, *Anatomy of the Village,* p. 31.
39. Sharp, T., Gibberd F., and Holford, W. G. (1953) 'The Design of Residential Areas', in *Design in Town and Village,* London: HMSO, Part II, p. 24.
40. Barry, Gerald. WORK25/44/A5/A4, 4 December 1948, National Archives, London.
41. This has similarities with Sharp's hypothetical scheme for a new town. See Pendlebury, 'The Urbanism of Thomas Sharp'.

42. Gibberd, F. (1953) *Town Design,* London: Architectural Press, p. 170.
43. Gutkind, E. A. (1951) 'The South Bank, a landmark in architecture: the discovery of the fourth dimension', *Architectural Design*, xxvi (December), pp. 349, 351.
44. Basil Taylor, 'English Art and the Picturesque', for BBC Third Programme, November 1953, 1 (of 3), transcript, p. 7.
45. Pevsner's Reith Lectures, published in 1956 as *The Englishness of English Art.*
46. Richards, J. M. (1951) 'Lansbury', *The Architectural Review*, 110 (December): 361–2.
47. Gibberd, F. (1976) 'Lansbury: The Live Architecture Exhibition', in Banham, M. and Hillier, B. (eds) *A Tonic to the Nation,* London: Thames & Hudson, p. 138.
48. Banham, R. (1955) 'The New Brutalism', *The Architectural Review,* 118 (December): 354–61.
49. Banham, R. (1968) 'Revenge of the Picturesque', in Summerson, J. (ed.) *Concerning Architecture,* London: Penguin.
50. Banham, R. (1976) 'Effeminate and Flimsy', in Banham and Hillier, *A Tonic to the Nation,* p. 193.

6

EVERYDAY UNAVOIDABLE MODERNIZATION AND THE IMAGE OF HELL

Visual planning in the writings of Nishiyama Uzō

Andrea Flores Urushima

Introduction

This chapter examines some elements of the planning theory of the Japanese planner and housing theorist Nishiyama Uzō (1911–94),* among which is the search for a reproducible standard – based on a generally shared way of living – and the search for a flexible and informative planning method. This theory evolved from an assessment of modernist ideas and the valuation of the visual experience in the process of urban change. Increasing distrust of the effectiveness of grandiose planning schemes born during the first half of the twentieth century has led practitioners and researchers to search for alternative approaches to the theme of city form. In this context Nishiyama's work stands counter to the mainstream of thought on city form in Japan, and his critical position regarding the work of his contemporaries enabled him to develop a substantial original output. This chapter will supplement the insufficient analyses of his ideas and influence to depict a more inclusive panorama of the anxieties of planning in mid-twentieth-century Japan.

Nishiyama Uzō, a former professor of Kyoto University and vice-president of the Architectural Institute of Japan in 1959, was a nationally influential figure in the fields of urban and housing theory with an extensive output of writing and plans that have a bearing on the education of a great number of currently active professionals in these fields. The Uzō Nishiyama Memorial Library, located in Kizugawa city in Kyoto Prefecture, maintains the invaluable archive of his whole life's output, which deserves further detailed analysis. This would help to develop a better understanding of the complex character of the mid-twentieth-century Japanese urban development that provided the basis for the miraculous economic renaissance and international integration of the country during the 1960s and 1970s.

This chapter will contextualize Nishiyama's original contribution to planning theory in Japan and will clarify the modernist background of his theory. Subsequently, the chapter will detail the analysis of two topics that are central to his planning theory: the search for a reproducible standard for quality of life, and the proposition of an informative planning method.

Formation of an original standpoint within the planning discipline in Japan

Like other planners such as the British Thomas Sharp, whose work is explored elsewhere in this volume, Nishiyama was interested in understanding industrial urban development and believed in the application of scientific principles to identify the qualities of a 'good' town in existing organic urban forms. At the same time, he had difficulties in putting his ideas into practice even though his influence as a reviewer spurred his professional career, such that he delivered important projects and became a powerful member of professional institutions. His isolated work process and his difficulty in implementing his ideas explain a lack of substantive research on his suggestions and influence. The scarcity of analysis of his work, in English[1] and in Japanese, might also be explained by an antipathy towards his personal character resulting from his criticism of governmental decisions and the work of his contemporaries. Nishiyama is commonly regarded in architectural and urban planning circles as a communist, although he was not a member of the Communist Party of Japan; and whereas the Marxist influence on his work is cited,[2] there still remains scope for a detailed and thoughtful essay on this theme. Finally, the criticism against the nationalistic tone of his discourse during the war[3] deserves a relative historical contextualization. During Nishiyama's youth in the late 1920s and 1930s, Marxism was an influential ideology across the whole spectrum of Japanese intellectual thought,[4] stimulating a debate centred around the figure of the emperor to overcome 'modern western individualism'.[5] During that period the conversion from Marxism to nationalism under official and police pressure, referred to as *tenkō* [conversion],[6] influenced the entire intellectual production of the time. Nishiyama followed a similar path to most of his contemporaries, and the changes in his discourse represent no conflict with the general intellectual changes of the period.

Nishiyama was born in a period of intense modernization and militarization in Japan.[7] Osaka developed into an industrial centre for supplying the military during the successive wars against China (1894–5) and Russia (1904–5): it had been known as 'Japan's kitchen' for its supply of essentials to the capital during the feudal period and was renamed during the modern period as the 'Manchester of the orient'.[8] The Nishiyama family owned an ironworking factory located in the Ajigawa area, on the same site as the family house (Fig. 6.1). The mouth of the Ajigawa River is historically famous as a strategic transportation node for transshipping merchandise from large ships to small boats, and it became a centre for shipbuilding and smokestack industries, such as iron and copper. The direct relation with the everyday life of the factory, located in the heart of industrial surroundings, put Nishiyama in contact with the workers' lives and allowed him to witness the consequences of the Japanese industrial revolution at first hand. Nishiyama's posthumously published book, a biography of his father, gives a dramatic yet poetic depiction of workers' lives in the area around the Ajigawa River during the end of the Tokugawa Feudal Regime in the mid-nineteenth century.[9]

His childhood might have stimulated his surveys of the ordinary life-style after his completion of his undergraduate studies in the Department of Architecture of Kyoto University (1930–3). His large-spectrum housing surveys of Japanese villages and towns encompassed from ancient to modern, rural to urban, wealthy to poor, including traditional historical town houses, houses of labourers and coal-miners, and impoverished ways of urban living. These surveys are documented in Nishiyama's books of the 1940s;[10] of these, *Dwelling of the Future: a Speech about Living Patterns* (1947)[11] (Fig. 6.2) received the Mainichi Publishing Cultural Award in March 1948 for its value as a record of the living conditions of common people.

FIGURE 6.1 Nishiyama's father's ironworking company in Osaka, the Manchester of the Orient, at the beginning of the twentieth century.
Nishiyama, U. and Nishiyama Uzō Kinensumai Machizukuri Bunko (1997) *Ajigawa monogatari: tekkō shokunin Unosuke to Meiji no Osaka* [A tale of Ajigawa River: Meiji Osaka and Unosuke, the ironworking artisan], Tokyo: Nihon keizai hyōronsha, p. 457.

Among his surveys, one interesting example is Hashima Island, located in Nagasaki Prefecture and known as *gunkanjima* [battleship island] due to its outer profile.[12] The island was previously a private undersea coal-mining island bought by the Mitsubishi Corporation in 1890.[13] At the turn of the twentieth century, as part of the remarkable Japanese industrial surge, the company sank two vertical shafts of about 200 metres, undertook land reclamation for industrial facilities and built high sea walls. In 1916, following a population rise, Mitsubishi built there the first reinforced concrete apartment block of significant size in Japan. By 1959 this tiny island of 6.3 hectares had reached a high density of 1,391 people per hectare in the residential district, inside buildings ranging from single-storey houses to concrete structures of up to nine floors. Food and most of the water were shipped from the mainland. Nishiyama and his students produced reports about the island's living conditions in 1954 and 1974[14] (Fig. 6.3), which stand among the few writings[15] about the island before the publication of detailed documentation in 1984.[16] Recently, in 2007 the island became an industrial patrimony preservation project supported by the Ministry of Economy, Trade and Industry.[17]

1 裏長屋の共同便所（江戸のウラダナ）

2 舊い町家の便所の型（京都の町家）

3 入口脇の便所（奈良・法蓮町）

4 裏側に突出した便所棟の列ぶ姿（大阪の長屋）

FIGURE 6.2a Survey of the common way of living in Japan. This series depicts the different disposition of external toilets in residential blocks: 1. Collective toilets in rear tenements (*uradana*). 2. Toilet type of old townhouses (*machiya* in Kyoto). 3. Toilet near to the entrance (at the Hōren-chō neighbourhood in Nara). 4. Line up appearance of toilet construction that stands out from the backyard (*nagaya* in Osaka).
Nishiyama, U. (1947) *Korekara no sumai: jūyōshiki no hanashi* [Dwelling of the future: a speech about living patterns], Tokyo: Sagami Shobō, pp. 154, 157, 158.

FIGURE 6.2b This figure is part of the series shown in figure 6.2a. It illustrates the collective toilets in rear tenements (*uranagaya*) and is a reconstruction of an imagined scene of one *uradana* of the Edo period drawn by Nishiyama.
Nishiyama, U. (1947) *Korekara no sumai: jūyōshiki no hanashi* [Dwelling of the future: a speech about living patterns], Tokyo: Sagami Shobō, pp. 154, 157, 158.

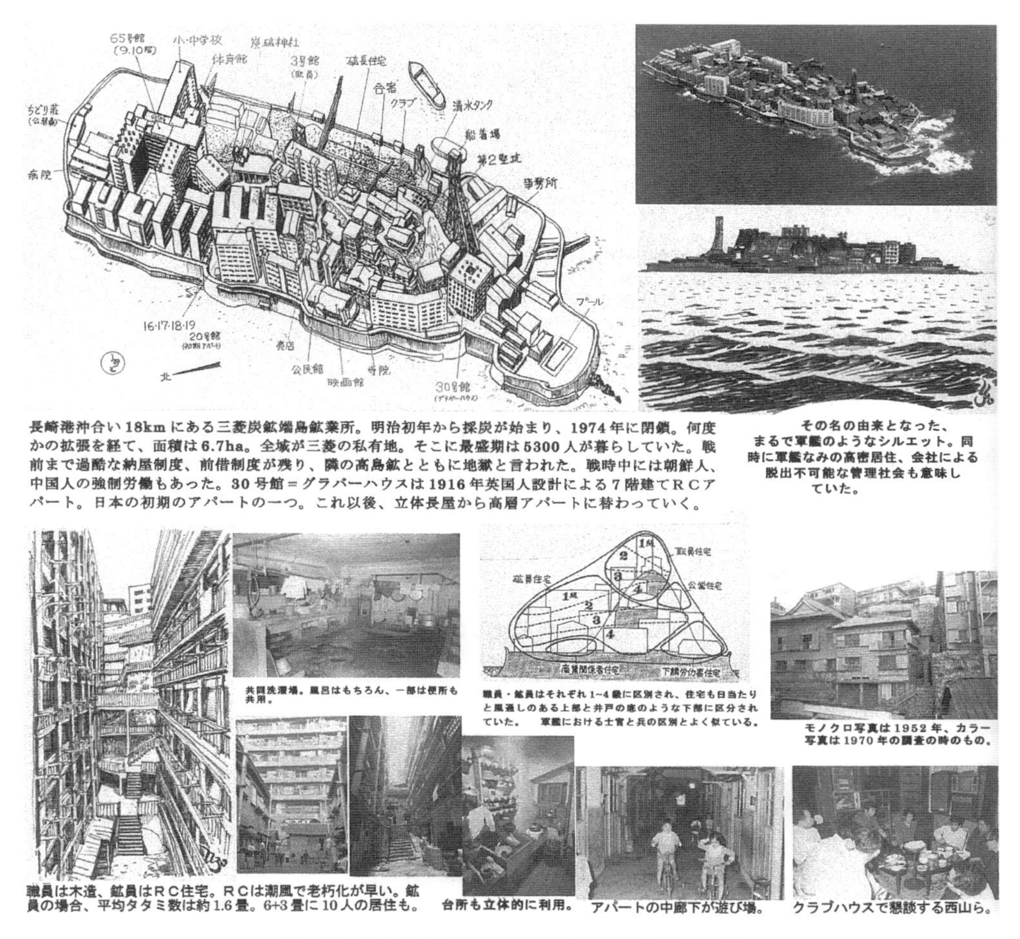

FIGURE 6.3 The Hashima island survey.
Nishiyama Uzō Kinensumai Machizukuri Bunko [The Nishiyama Uzō Memorial Library] (2000) *Nishiyama Uzō to sono jidai* [Nishiyama Uzō and his time], Kidzuchō (Kyoto): Nishiyama Uzō kinen sumai machizukuri bunko, p. 84.

This example illustrates Nishiyama's pioneering interest in the comprehension of Japanese modern industrial development through housing research, and his original standpoint in terms of a sociological perspective on the analysis of urban settlements. Empirical sociology based on surveys began to appear in Japan in the 1920s in parallel to discussions about the search for the 'uniqueness of Japan' and as part of discussions about overcoming universal western canons. To this end, surveys were mainly undertaken in rural areas, whereas the modern urban phenomenon was largely ignored. Urban sociology and surveys of modern cities mostly developed during the postwar period with a few earlier exceptions, such as the work of the sociologist Okui Fukutarō. Similarly, in disciplines such as architecture and urban planning some early surveys were undertaken specially in rural areas with rare analyses of urban examples. One of the first architects who began to survey the everyday life-style in Japan in the 1920s was Kon

Wajirō, whereas the prewar period is short of relevant work based on surveys of the everyday modern life-style in urban settlements. In this context, Nishiyama's surveys of several different living contexts, both urban and rural, especially the surveys which document the direct impact of industrialization, serve as a valuable reference for the unprecedented development of an empirically based planning discipline in Japan.

In addition, the Japanese architectural historian Funo Shūji noted that the early period of Nishiyama's writing, from plans to the management and production of architecture, brought an innovative perspective to the Japanese literature. Funo asserts that Nishiyama's writings changed the focus from architectural theory to housing analysis, from when he started to develop the successive surveys on the way of life of ordinary people.[18] In parallel with the housing surveys, Nishiyama stepped up to the scale of urban theory, and this shift can be found in his collection of texts written between 1942 and 1968.[19] Nishiyama began to address urban planning analysis in texts from the 1940s. His 1950s output is largely related to surveys of the ways of life of ordinary people, and during the 1960s, his urban theory reached a proactive form of sketched plans.[20] His analysis of large scope – in terms of its spatial scale and sociological contents – aimed at the fundamental development of a planning theory to adequately voice the specific anxieties of Japanese modernization against universalistic models.

Laying the theoretical foundations: modernist influence reassessed

From another perspective, it is relevant to highlight Nishiyama's solid intellectual foundation on modernism, especially the European influence he received in the Department of Architecture. The first Japanese modernist architectural movement, *Bunriha kenchikukai* [Secessionist Architectural Group],[21] inspired by the Viennese Secession of 1920–8, greatly influenced all of Nishiyama's generation.

In his *Notes on Architectural History* (1948),[22] a compilation of texts from 1933–7, Nishiyama defended the need for a better understanding of the peculiar characteristics of the architectural discipline in Japan. He stressed four main aspects of this discipline, the first three of which are the living aspect (the study of plans), the physical aspect (the study of structural issues) and the productive aspect (the study of executive, management and administrative issues). He critically asserted that the artistic aspect, the fourth element, became restricted to the study of architectural history and styles of the past, especially those used by a 'bourgeois' ruling elite in western countries. Against this fallacious historicism, Nishiyama advocated an architectural history locally rooted in historical forms, materials and ways of living. This assertion reflects a subtle influence of the *Bunriha* Secessionist discourse, with a mention of the *Bunriha* as the driver of many of the founders of the Japanese modernist tradition of the 1920s, such as *Sōusha* [The creation of the universal society] of Okamura Bunzō in 1923, *Ratō* [Naked fight] of Kishida Hidetō in 1924 or *Meteōru* [Meteor] of Imai Kenji and Satō Takeo in 1924.[23]

His text also highlights the theme of 'locality' in opposition to 'international' (style) in the writings of the *Nihon intānashonaru kenchikukai* [International Architectural Association of Japan], the only Kyoto-based pioneer modernist group, centred around the figures of Ueno Isaburō and Motono Seigo in 1927.[24] In the historical context of the formation of the 'locality' concept, Nishiyama cited the housing debate at the second 'International New Architecture Meeting' in Frankfurt am Main in 1927.[25] This is a reference to the *Neues Bauen* discussions about *Existenzminimum* standards and the 1924–30 series of plans for the New Frankfurt headed by the German architect and urban planner Ernst May. The mention of a second meeting is a probable reference to the meetings of the group of architects originally gathered into the *Novembergruppe,* which subsequently changed its name to the 'Circle of Ten' in 1924 and finally established 'The Ring' in 1925. This group included Walter Gropius, Mies van der Rohe and Peter Behrens among others.[26] Nishiyama's 1948 book repeatedly refers to Gropius, May, Behrens and the Russian architect Alexander Klein,[27] demonstrating the influence of the German *Werkbund* on Nishiyama's emerging ideas.

Nishiyama's ability to read German-language written materials is known. However, his comprehension level of English written materials, especially during his young period, is unclear. Thus, it is possible that the German early modernist discussions on collective housing served to introduce Nishiyama to the British early modern debate. Nishiyama reproduces images of the planning of neighbourhood units from Thomas Adams's 1934 book *The Design of Residential Areas*. In the same text two images appear without a source citation: a plan by Robert Whitten and Gordon Culham, and another plan signed by the Garden Community Committee[28] (Fig. 6.4). Nishiyama also possessed the books *The Anatomy of the Village* (1946) and the Japanese translation of *Town and Townscape* (1968) of Thomas Sharp – who had worked with Adams. Among the earliest foreign publications in his personal collection were two books about the British biologist and urbanist Patrick Geddes. The oldest English book in his personal collection dates back to 1945 and was preceded by German and Russian books.[29]

Later, in 1974, as documented in his retirement speech, Nishiyama evaluated the contradictions of the 'functionalist and rationalist' modernist foundation of his theory. He cited the participation of Otto Wagner in the Viennese Secession as one revolutionary precursor of the modern movement. Among the negative aspects of the modernist development, he categorically criticized some formalist currents of European modernism for freezing the architectural revolution inside white geometrical forms of dogmatic 'ostentation' [*kyoshoku* in Japanese]. Nishiyama criticized the exaggerated aesthetic preoccupation with the artistic value of an architecture based on the inspired creation of talented 'geniuses'. He associated these formalist currents with the undergraduate lectures on residential design he had received from one of the

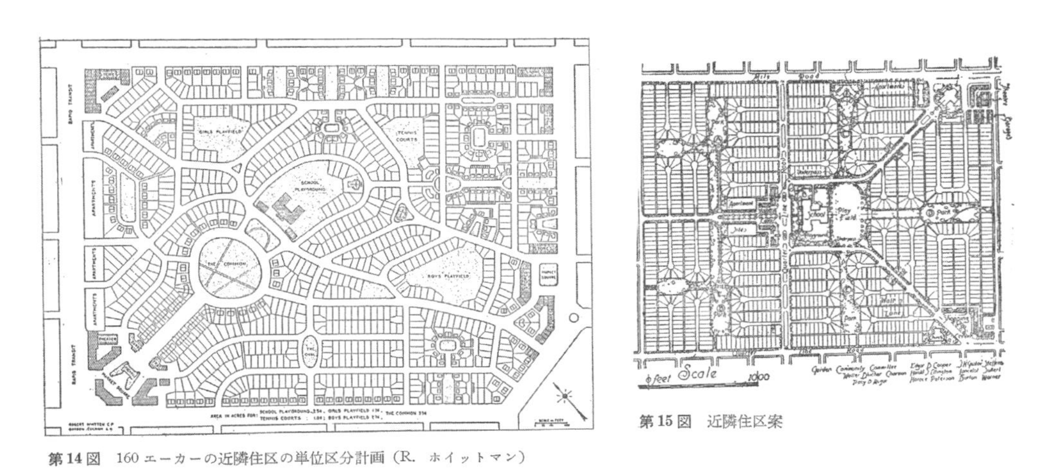

FIGURE 6.4 A plan of Robert Whitten and Gordon Culham (left) and a plan signed by the Garden Community Committee (right).
Nishiyama, U. (1968) *Chiiki kūkanron* [Regional space theory], Tokyo: Keisō Shobō, pp. 50–1.

pioneers of Japanese modern architecture, Takeda Goichi, in which the students were obliged to learn how to design houses with billiard rooms and wine cellars, among other features, and what the effects on wine were of keeping beer and wine together in the same space. This lifestyle portrayal was unprecedented for Nishiyama and was far from his childhood experience of industrial reality in Japan: it came with his discovery of the meaning of the word 'bourgeois'.[30]

In contrast, he praised in the early origins of modernism the claim for social responsibility, the comprehensive new approach encompassing tasks ranging from design to planning, and the change in point of view from formal design to the organization and meaning of space itself. Moreover, against the inspiration of 'geniuses', modernism also nominated objective measurable elements of architectural production to equip 'normal' people who wished to become architects. This comprised the search for a scientific method for the discipline of architecture exemplified in Klein's investigations and the search for the objective measurement of the luminosity, acoustic levels and temperature, for example, inherent in architectural forms and materials.[31]

The search for a standard for quality of life

In 1942 Nishiyama wrote an article entitled *Seikatsu kichi no kōzō* [The structure of a life base][32] in an attempt to investigate the elements that define quality of life. At that time he was working in the *Jūtaku eidan* [Housing Corporation], a state agency created during wartime to produce housing on a large scale for the increasing numbers of 'industrial soldiers' flowing to defence plants and urban factories.[33] One of the major discussions of the period was how to define standard patterns of the minimum quality for housing, a difficult question for Nishiyama because he considered that the definition of the ventilation area or floor plan area was variable and related to cultural factors, public morals and welfare. He considered that economic coercion and the ruling state's need to control human power during the general wartime mobilization pressed architects to the complex task of defining a total number for quality of life and, at the same time, creating a method to quantify it.[34]

The 1942 text affirmed that several rules to generate a medium level of residential quality had been defined worldwide since the nineteenth century through the regulation of architectural construction. However, he argued that he was unavoidably pushed to tackle that task at a moment when the central government was pressing architects to 'bear the weight' of defining proper national and social standards for housing. To fulfil this task he suggested the importance of undertaking a comprehensive investigation about quality of life, including all aspects of life, instead of a separate evaluation of the various aspects of the housing theme. In the text he used the expression *seikatsu rin'ne zentai,* where *seikatsu* means 'life'; *rin'ne* is a Buddhist concept meaning *samsara,* or 'the endless cycle of death and birth'; and *zentai* means 'whole'. The expression suggests the inclusion of transcendental or non-rational elements of life.[35]

Nishiyama described the process of large-city formation with the statement that the modern patchwork-like type of urban development was ruled by the logic of capitalism rather than resulting from a unifying plan of improvement. He also affirmed that the people able to consider an ordered, rationalized and improved new life should seek to move towards this, for otherwise a new life would result from adaptation to the needs of daily life.

This text culminated in a critical analysis of propositions to control urban growth that had been advanced since the end of the nineteenth century. Nishiyama gave a positive evaluation of the Garden City theory (Fig. 6.5), while he criticized Letchworth and Welwyn because of the unusual enthusiasm of enterprisers who owned large tracts of land under circumstances that only an advanced capitalist country like England permitted. He affirmed that such a difficult enterprise would never progress without a perfect agreement on profits made by the builder, the land owner, the enterpriser and the banker, among others, under the circumstances of a free market. Moreover, he added that the Garden City movement was not a precursor of urban planning in England but a complement to an existing Town Planning regulation, such as the 1909 Act. These considerations suggest that the Garden City ideas as implemented in the form of Letchworth and Welwyn were not a large-scale solution to the urban problem but rather a specific type of intervention that would be difficult to implement in Japan. He added that the idea of garden suburbs and satellite cities flourishing under the Garden City influence should be used with caution since the long-term consequences of that type of planning experiments were still unclear.

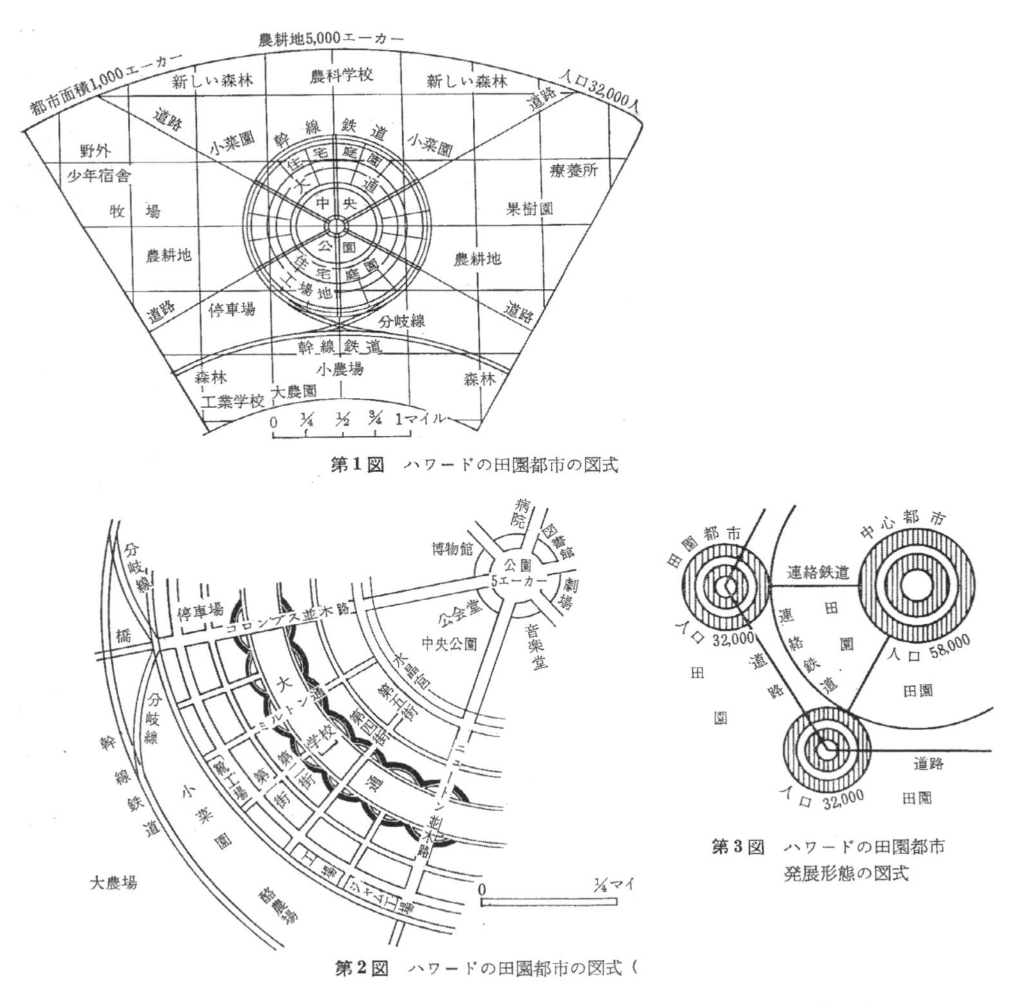

FIGURE 6.5 In Nishiyama's article originally written in 1942, the translated sketches shown in the text include one sketch that does not come directly from Howard's book but is a simplified scheme of Howard's original network concept. It is possible that Nishiyama decided to do his own translation of the original source.
Nishiyama, U. (1973) *Chiiki kūkanron* [Regional space theory] (4th ed.), Tokyo: Keisō Shobō, pp. 50–1.

The plans he considered to be nothing more than formal architectural reform of the city, devoid of new propositions for the urban way of living, are gathered in a section entitled 'ideal cities', which includes the 1911 prize plan of Canberra, together with the 1922 Ville Contemporaine of Le Corbusier. In this section Nishiyama criticized the 1927 *Hochhausstadt* [High-rise city plan] of Ludwig Hilberseimer as a plan that formally separated the existing infrastructure of the spontaneous city in a manner that is inefficient in terms of the reorganization of city functions and everyday life.[36]

Moreover, Nishiyama considered the Garden City-influenced model of the Spaniard Arturo Soria y Mata to be formally related to the Russian idea of the linear city presented in the work of Nikolai Miliutin. However, he affirmed that these formally similar models had dissimilar originating theories due to the distinct location of production facilities. He termed the Spanish model the *Obijō toshi* [belt city] instead of the more usual *Taijō toshi* [linear city] because in the Spanish model the production facilities were located outside the linear development. In contrast, in the Russian model the housing, green areas and public facilities developed in a continuous linear pattern in parallel to the production facilities, which were strategically disposed according to the necessity of a production line and well served by a transportation system of railways and roads. This section closed with the assertion that the Russian model was the nearest to an 'ideal' urban form but that the lack of qualified and experienced planners obstructed the satisfactory development of the model in the many trials built in the Soviet Union.

The recollection finished with a neutral overview providing no critical comment on the plans of the 1939 *Die neue Stadt* [The new city] of Gottfried Feder (Fig. 6.6). Nishiyama considered the definition of the scale and type of urban institutions of these plans to be well grounded in existing urban conditions. Moreover, he asserted that Feder's plan for the Russian town of Kirs was assembled in an organic way where the institutions correlated with the complete circle of life, *samsara* or *rin'ne,* in groupings defined by the categories of daily, weekend and monthly use. This differed from the Russian linear cities, in that Feder's plan aimed to a great extent at a revival of small medieval cities.[37] Hein has highlighted Nishiyama's interest in the ideas of Feder, among which is the intention to promote a 'new art of city planning' that included the social structure of the inhabitants in the search for an aesthetic element representative of a local autonomy.[38]

This historical recollection forms what Nishiyama calls 'precedents' or the references for the development of his investigation about the planning of the way of living. The preoccupation with the spatial materialization of a 'quality of life', from the scale of a house's interior to an extended regional scale, was a permanent theme of his research. For him the most important element of the production of space at all scales rested in the essential question of the 'way of life' and the difficulty of measuring life in numerical form. His surveys of the ways of life of ordinary people occurred in response to his distrust of the formalist currents of modernism and his recognition that the general mass of ordinary people were dissatisfied with the outcomes of functionalism. Through his surveys Nishiyama realized that the habits and traditions continuously reassessed in contemporary transformations underscored a logic of practice, which ultimately shaped the way of life of ordinary people. Moreover, the quality of a house was directly related to the quality of the services at an extended regional scale, such as transportation or public institutions. Finally, he admitted that it was impossible to draw a regional plan in the same way as in architecture, because of the collective scale and time frame of urban transformations. Moreover, these transformations were a result of the conflicting relationship between space and life, mediated by the permanent exertion of a power struggle between social

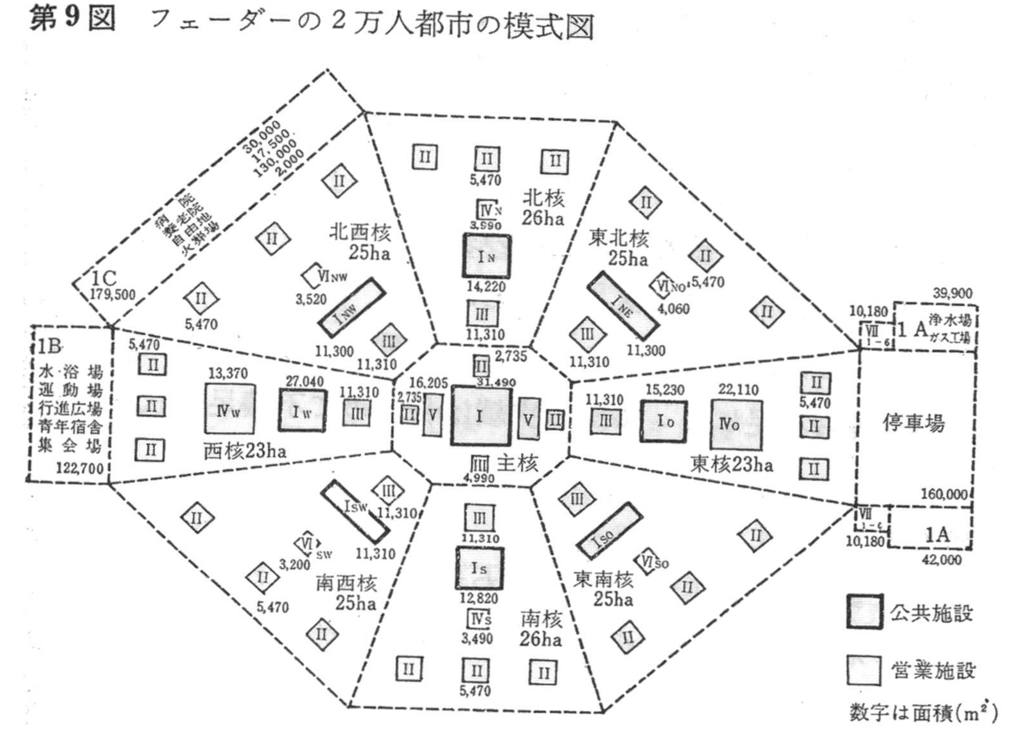

FIGURE 6.6 Feder's city for 20,000 people.
Nishiyama, U. (1968) *Chiiki kūkanron* [Regional space theory], Tokyo: Keisō Shobō, p. 37.

groups. He suggested that plans and prospective visions were a useful intermediary to voice the needs of common people against the hegemonic control over city form held by those with economic and political power.[39]

The search for a planning method: Image Planning and the vision of hell

In order to use plans as an impartial instrument of urban renovation, during the 1960s Nishiyama proposed to put in practice the 'Image Planning' method of producing synthetic master plans containing detailed investigations about existing cities and their formation to orientate future intervention.[40] The origins of the concept of Image Planning go back to a text of 1949[41] about the instrumental use of imagined ideals to reach concrete achievements. The text considers the historical evolution of Japanese cities under the oppressive control of a ruling elite which progressively removed the 'creative' powers of ordinary people. It came to the assumption that the same oppressive ruling elites progressed an industrial revolution that led to the consolidation of a generalized poor quality of life. On the basis of this paradoxical evolution – the development of technology to the detriment of social progress – Nishiyama considered the contradiction existing between concrete form and social demands, recalling the enormous distance between the modernist images of the purity and cleanliness of 'Everytown' – the ideal future urban improvement shown in the 1936 film *Things to Come* – and the reality of cities in interwar Japan.[42] Moreover, he acknowledged that this distance between what is dreamed and what is implemented was a result of the difficulty of large-scale vision in the common people's imagination:

> . . . Bees make complex beehives which astonish anyone by means of a dexterity that any architect cannot reach. But they only follow their instinct to make these. As opposed to this, in the human case, before making a house, even a bad architect clearly visualizes inside his head what will be built. . . . People usually first visualize a big picture in their heads of what they can make by themselves. All the constructions and all the public facilities that comprise the city have always been created in this way. But have we imagined in our thinking 'Let us create this kind of city' like the one we are living now? All these city parts have been created inside people's heads by a plan visualized at once, but the huge modern city is composed by the wholeness of the parts and was never imagined in our heads consciously before.
>
> *(Translated by the author from the original in Japanese)*[43]

Nishiyama contrasts bees' instinctive way of building their nests to the human visionary way of building cities. He considers that the visionary precondition exists at any scale: from one garden, to one house, to several buildings. However, because cities are the composition of different individual visions – resulting from different social standpoints – it is very difficult to implement one final form of a whole city that can adequately respond to the social demands existent in every partial individual vision.

In this context, Nishiyama believed that regional and urban planning served to resolve the distance between the partial imagination of solutions for everyday needs and the final form of a whole urban ensemble (Fig. 6.7). At the same time, planning required the imagination of an ideal urban form to unify the scattered parts of the city that had grown out of everyday needs. Nishiyama affirmed that an ideal urban form should aim at a resolution of the diverse contradictions of existing cities and should also be able to forecast the new problems brought by the newly proposed solutions. He considered that all of the ideal cities drawn in the past had been born from the necessity to solve the evils contained in existing cities; in contrast, a useful model for real cities should be capable of raising debate about existing problems while being flexible to accommodate change over time. Illustrating these considerations are, among others, depictions of the closed ideal cities of the Renaissance period of Europe, the plans of Albrecht Dürer, the 1849 'Victoria' plan of James S. Buckingham and the 1922 *Ville Contemporaine* plan of Le Corbusier.

Nishiyama stated that the fundamental difference between the prospective vision of regional planning and futurology, a then-popular discipline heavily based in the computer technology of the 1970s, was that 'the problem of real regional space planning is that actual measures should receive the feedback of an estimation of what will happen in a new future and a clear idea of the possible dangers of the general previewed plan'.[44] He emphasized the tendency to neglect the fact that plans and decisions made to solve existing problems could possibly bring forward new unforeseen problems (Fig. 6.8).

He defended the use of *jigokue* [images of hell] to alert people: science should bravely extract the problematic substance of any proposition in order to turn the plan into an instrument of debate. For him the most appropriate plan was one that synthesized an interdisciplinary proposition that accounted for a scientific consideration of positive and negative results, by means of drawing images, collecting information and writing reports. He suggested that the visual element of plans was often used solely to draw positive views for the future. In contrast, Nishiyama

FIGURE 6.7 Non-planned city of 1955.
Photo: Miki Jun. Shigemori, K. (ed.) (1987) *Nihon shashin zenshū 7: toshi no kōkei* [*The complete history of Japanese photography: the spectacle of cities*], Tokyo: Shōgakukan, p. 55.

FIGURE 6.8 Planned city of 1964, Matsubara Danchi.
Photo: Fukase Masahisa. Nagano, S., Īzawa, K. and Kinoshita, N. (eds) (1998) *Nihon no shashinka 34* [*Japanese photographers 34*], Tokyo: Iwanami shoten, p. 12.

pushed intellectuals and scientists to use drawing techniques to imagine and show the hell that a proposition might bring, in a way that the image itself remained a neutral tool for citizens to discuss and imagine by themselves solutions to the problems of their everyday lives.

The quest for visual investigation in Nishiyama's work

One important feature of Nishiyama's work is the interest in objectifying the formal elements of spontaneously created settlements in order to reproduce them through the intentional action of a planner. The visual experience is important both for the sake of observing existing cities and also for supporting the imagination of new urban structures. For Nishiyama the visual experience existed in the imagination of people who were motivated to intervene in the improvement of their surroundings. As one inevitably immersed in the new ocularcentric era,[45] Nishiyama demonstrated the modern approach of his theory in his emphasis on the powerful influence of images, especially his consideration that the subjective mentally visualized image forms the original foundation of any human action.

The development of new media and their influence in Nishiyama's planning conceptions are readily recognizable. The impact of the unlimited possibilities of the visually realistic sensations of the moving image appears in his powerful criticisms of 'Everytown'. Moreover, the influence of advertising in the process of citizens' consumption of city making is apparent in his proposition. For Nishiyama, the city is a commodified physical artefact to be negotiated and consumed with reference to the imagined desires of people. This view echoes the German *Werkbund* discussions about quality in an industrial society and well-informed consumption.[46] Within Nishiyama's interpretation the city can be considered as a commodity, the citizen as a consumer, the architect as a producer and the plan as a shop-window or salesperson. The exposure of the possible failures of a plan through the vision of images of hell brings quality to the consumption of the city by inviting citizen participation in the process of city making.

Nishiyama advocated the use of an impartial visual representation to provoke people into a debate. His attempt to encompass the wholeness of a vision, in its positive and negative aspects, implies his acknowledgement that the plan is an empty frame. Depending on how this frame dislocates towards more optimistic or pessimistic visions, an entire non-visualized universe gains concreteness through visual representation. This interpretation suggests a similarity to the understanding of 'gaze' in the interpretation of the Japanese philosopher of the Kyoto School Nishitani Keiji.[47] In Nishitani's approach to 'gaze', *sun'yata,* or the 'blankness', encompasses the wholeness of the universe that exists before and behind a spectator's field of vision. Even if the subjective point of view is non-existent and the objects are not made visually concrete, the universe still exists in the form of an empty vision. In contrast to the French philosopher Jean-Paul Sartre, who considers that the object can exist only through subjective visual materialization, Nishitani asserts that the object exists beyond the subjective desire active in the move of the subjective eye.

Nishitani's *sun'yata* interpretation serves as a reference to explain Nishiyama's attempt to incorporate the image of hell [*jigokue*] into an imagined wholeness of a 'way of living'. According to Nishiyama, what does not appear in the drawings of a plan should not be considered non-existent, but rather should be visually materialized through the dislocation of the active frame that the drawn plan provides. Modern western planning drawings are commonly used to show desired positive and appraisable images of what the city will become. However, Nishiyama emphasized the unimagined or undesired consequences latent in any concrete intervention. By doing this he drove architects and urban planners to dislocate their viewpoint to what is out of sight or to the blankness of the global existence.

Conclusions

Until the 1920s there were few departments of architecture inside universities, and only a recent history of the establishment of professional architectural offices, beginning from the early twentieth century. At this time, alongside a discussion of the meaning and content of a Japanese architectural theory and history, the nascent architectural discipline in Japan readily incorporated the effervescent modernist debate from Europe. The Japanese national identity debate and the question of tradition and 'locality' versus European-centred 'internationalization' became relevant in that context. Against a fallacious historicism and with a critical opinion opposed to the dogmatic universal value of some currents of international modernism, Nishiyama advocated the scientific objectification of the everyday life-style in Japan. The understanding of local problems and the search for potential solutions within the peculiar historical circumstances of Japan lay behind Nishiyama's surveys of the Japanese 'way of life'. The historical development of Japanese settlements and their contradictions have filtered the particular interpretation of his vision for planning. On the one hand, he admitted that all concrete interventions within a limited practical scale were mediated by imagined visions. On the other hand, he emphasized that modernization brought complexity to the visual imagination of large-scale ensembles, which makes modern planning an indispensable discipline. He defended the idea of planning new settlements by means of a future vision encompassing the wholeness of the human 'way of living' with consideration given to the needs of everyday activities, as much as to the greater needs of the maintenance of a larger urban ensemble. At the same time, he defended the creation of a planning methodology supported by the wholeness of a future vision containing both the positive and negative consequences of any intervention. For Nishiyama, urban planners had an informative role to support the community and government in the search for solutions to the contradictions of the modernization of everyday life.

Acknowledgement

The author would like to express his gratitude to the members of the Uzō Nishiyama Memorial Library, with special mention to Professor Hirohara Moriaki, for the kind support for his research and for permission to reproduce the images used in this chapter.

Notes

Translated book and journal titles in italics (i.e. as per Japanese original)

* This paper follows the Japanese convention of placing family names before given names.

1. C. Hein, 'Visionary plans and planners: Japanese traditions and western influence', in N. Fiévé and P. Waley (eds) *Japanese Capitals in Historical Perspective: Place, Power and Memory,* London: Routledge Curzon, 2003; A. Flores Urushima, 'Genesis and culmination of Uzō Nishiyama's proposal of a "model core of a future city" for the Expo 70 site (1960–73)', *Planning Perspectives,* 2007, vol. 22, 391–416.
2. T. Muramatsu, *Nihon kenchikuka sanmyaku* [Mountain ridge of Japanese architects], Tokyo: Kagoshima shuppankai, 1985, p. 102; S. Funo, *Kokka, yōshiki, tekunorojī: kenchiku no Shōwa* [State, style, technology: the architecture of the Shōwa period], Tokyo: Shōkokusha, 1998, p. 104.
3. The Japanese historian of architecture Fujimori Terunobu discusses the reasons behind the participation of architects, including Tange and Nishiyama, in one very nationalistic project of the Great East Asian Co-Prosperity Sphere Memorial Competition (*Daitōa* competition, 1942). He depicts an active nationalistic portrait of Nishiyama during the war. Moreover, he suggests that Nishiyama's posterior reassessment of his nationalistic period lacked self-criticism. For the discussion about the *Daitōa* competition see K. Tange and T. Fujimori, *Kenzō Tange,* Tokyo: Shinkenchikusha, 2002,

pp. 80–2. For the discussion about the nationalistic period and the architects see ibid., pp. 104–9. Specifically about Nishiyama, see ibid., p. 108.

4. A.E. Barshay, *The Social Sciences in Modern Japan: the Marxian and Modernist Traditions,* Berkeley: University of California Press, 2004, p. 55; for avant-garde artistic movements influenced by Marxism see Y. Abe, 'Un bref moment de l'histoire', in Centre Georges Pompidou (ed.) *Japon des avant gardes 1910–1970,* Paris: Éditions du Centre Pompidou, 1986, pp. 91–3.
5. O. Eiji, 'Postwar Japanese intellectuals' changing perspectives on Asia and modernity', in S. Saaler and J.V. Koschmann (eds) *Pan-Asianism in Modern Japanese History: Colonialism, Regionalism and Borders,* London: Routledge, 2007; K. Karatani, 'Overcoming modernity', in R.F. Calichman (ed.) *Contemporary Japanese Thought,* New York: Columbia University Press, 2005, pp. 102–18.
6. For example P.G. Steinhoff, *Tenkō: Ideology and Societal Integration in Prewar Japan,* New York: Garland, 1991; A. Itō, *Tenkō to tennōsei: Nihon kyōsanshugi undō no senkyūhyakusanjū nendai* [Conversion and the emperor system: the Japanese communist movement of the 1930s], Tokyo: Keisō shobō, 1995.
7. Nishiyama Uzō Kinensumai Machizukuri Bunko [The Uzō Nishiyama Memorial Library], *Nishiyama Uzō to sono jidai* [Nishiyama Uzō and his time], Kizuchō (Kyoto): Nishiyama Uzō Kinensumai Machizukuri Bunko, 2000.
8. Urban Engineering Information Center Osaka Municipal Office, *Osaka and its Technology,* pp. 36, 37, Osaka: Osaka Municipal Government, 2000, pp. 36–9; J.E. Hanes, *The City as Subject: Seki Hajime and the Reinvention of Modern Osaka,* Berkeley: University of California Press, 2002, p. 173.
9. U. Nishiyama and Nishiyama Uzō Kinensumai Machizukuri Bunko [The Uzō Nishiyama Memorial Library], *Ajigawa monogatari: tekkō shokunin Unosuke to Meiji no Osaka* [A tale of Ajigawa River: Meiji Osaka and Unosuke, the ironworking artisan], Tokyo: Nihon keizai hyōronsha, 1997.
10. U. Nishiyama, *Jūtaku mondai* [The housing problem], Tokyo: Sagami shobō, 1942; U. Nishiyama, *Kokumin jūkyo ronkō* [Citizen's dwelling essay], Tokyo: Itō shoten, 1944.
11. U. Nishiyama, *Korekara no sumai: jūyōshiki no hanashi* [Dwelling of the future: a speech about living patterns], Tokyo: Sagami shobō, 1947.
12. Nishiyama Uzō Kinensumai Machizukuri Bunko, op. cit., p. 84.
13. B. Burke-Gaffney, 'Hashima the ghost island', *Crossroads: a Journal of Nagasaki History and Culture,* 1996, vol. 4, 33–52, at pp. 33–5.
14. U. Nishiyama and M. Ōgida, 'Gunkanjima no seikatsu: Nagasaki kōgai, Mitsubishi hashima tankō no kengakuki' [The life in the Battleship island: field trip memoir of Mitsubishi's coal mine Hashima, the outside harbour of Nagasaki], *Jūtaku kenkyū* [Housing research], 1954, vol. 1, no. 4, 41–51; T. Katayose and H. Fujinaga, 'Gunkanjima no seikatsu kankyō: kenchiku keizai jūtaku mondai' [The housing question, architectural economics and living environment of the Battleship island], *Taikaigakujutsu kōenkōgaishū keikakukei – Summaries of Technical Papers of Annual Meeting: Planning,* no. 49, Tokyo: Architectural Institute of Japan, 1974, pp. 1395–6.
15. K. Fudō, 'Sekai ichi no jinkō mitsudo, midori naki shima, Hashima' [Hashima, the island without green: the highest population density of the world], *Kokusai bunka gahō* [International cultural pictorial], 1951, vol. 3, 4–9; W. Hitose, 'Hashima fūkei' [The scenery of Hashima], *Chiiki* [Area], 1952, vol. 1, no. 4, pp. 36–9.
16. Y. Akui and H. Shiga, *Gunkanjima jissoku chōsa shiryōshū Taishō Shōwa shoki no kindai kenchikugun jishōteki kenkyū* [Collection of source information from a survey of the Battleship island: a case study of the modern architectural assemblage from the Taishō and Shōwa era], Tokyo: Tokyo denki daigaku shuppankyoku, 1984, p. 21.
17. NPO hōjin Gunkanjima wo sekai isan ni surukai [NPO the way to world heritage Gunkanjima], *Gunkanjima wo sekai isan ni surukai kōshiki WEB* [The way to world heritage Gunkanjima official website], Nagasaki: NPO hōjin Gunkanjima wo sekai isan ni surukai, available online at <www. gunkanjima-wh.com/> (accessed 30 March 2010).
18. S. Funo, op. cit., pp. 105–6.
19. U. Nishiyama, *Chiiki kūkanron* [Regional space theory], Tokyo: Keisō shobō, 1968.
20. A. Flores Urushima, op. cit.
21. The most recent account in Japanese is D. Amanai, 'Bunriha kenchikukai kessei no rironteki haikei: shoki Nihon kenchikukai ni okeru "geijutsu" to "hyōgen"' [The founding of *Bunriha kenchikukai:* 'art' and 'expression' in early Japanese architectural circles, 1888–1920], *Bigaku* [Aesthetics], 2007, vol. 57, no. 4, 69–82. In English, see J.M. Reynolds, 'The *Bunriha* and the problem of "tradition" for modernist architecture in Japan 1920–1928', in S.A. Minichiello (ed.) *Japan's Competing Modernities,* Honolulu: University of Hawaii Press, 1998, pp. 228–46, and K.T. Oshima, *International Architecture in Interwar Japan: Constructing Kokusai Kenchiku,* Seattle: University of Washington Press, 2009.

22. U. Nishiyama, *Kenchikushi nōto* [Notes on architectural history], Tokyo: Sagami shobō, 1948.
23. Ibid., pp. 4–6, 191–211.
24. An analysis of the 'locality' concept is found in K. Kasahara, 'Nihon intānashonaru kenchikukai ni okeru Itō Masabumi no katsudō to kenchiku rinen ni tsuite – A study on works, activities and architectural ideas of Masabumi Itō in the International Architectural Association of Japan', *Nihon kenchikugakkai keikakukei ronbunshū – Journal of Architecture, Planning and Environmental Engineering,* 2003, no. 566, pp. 153–9.
25. U. Nishiyama, 1948, op. cit., p. 193.
26. D. Sharp, *Modern Architecture and Expressionism,* London: Longmans, Green and Co., 1966, pp. 63–4, 82.
27. In 1932 Nishiyama wrote about Klein's floor plan design research for public housing in Germany. See C. Hein, 'Nishiyama Uzō to Nihon ni okeru seiyō riron no denpa' [Nishiyama Uzō and the spread of western concepts in Japan], *10+1 Ten plus One,* 2000, no. 20, pp. 144–5.
28. U. Nishiyama, 1968, op. cit., pp. 50–3.
29. The two earliest foreign-language books in Nishiyama's personal collection were Deutscher Werkbund, *Bau und Wohnung: Die Bauten der Weißenhofsiedlung in Stuttgart errichtet 1927 nach Vorschlägen des Deutschen Werkbundes im Auftrag der Stadt Stuttgart und im Rahmen der Werkbundausstellung 'Die Wohnung',* Stuttgart: Akad. Verlag Dr. Fr. Wedekind & Co., 1927; and E. Nagel, *Grundlegende Wertzahlen über Wohndichte und Besiedelungsdichte im Städtebau,* Munich: Georg D. W. Callwey, 1927.
30. Nishiyama Uzō sensei taikan kinen jigyōkai [Commemorative council of the retirement ceremony of Prof. Nishiyama Uzō], *Nishiyama Uzō sensei taikan kinen shūroku* [Memorial compilation of Professor Nishiyama's retirement ceremony], Kyoto: Nishiyama Uzō sensei taikan kinen jigyōkai, 1974, pp. 11–15.
31. Ibid., pp. 12–13.
32. *Seikatsu* means 'life' or 'living', *kozo* means 'structure', and *kichi* means 'base', usually used to refer to a military/naval/space base. However, instead of composing this title with words such as 'military/naval/space', Nishiyama plays with the words and uses 'life'. The first Chinese character of the word *kichi* means 'foundation/basis' and is also used in words such as *kiso* (fundament or foundation of a theory), *kijun* (standard or norm) and *kiban* (general base or foundation of a civilization, of a book, etc). Nishiyama probably wanted to keep the ambiguity of the word 'BASE' for the title (both an implicit reference to a military base and to 'base' in the sense of a standard or foundation).
33. G. Clancey, 'Designing a home for the Yamato Minzoku: race, housing and modernity in wartime Japan', *Asian Studies Review,* 2005, vol. 29, no. 2, pp. 123–41.
34. S. Toyokawa, 'Tange kenkyūshitsu ni okeru jūtaku keizairon: sōryoku senka no saiseisan to seikeihi' [The housing economic theory in Tange's Laboratory: reproductive and standard cost of living under the general war mobilization], *Nihon kenchikugakkai keikakukei ronbunshū – Journal of Architecture and Planning: Transactions of the Architectural Institute of Japan,* 2006, no. 609, p. 176.
35. U. Nishiyama, 1968, op. cit., p. 21.
36. Ibid., p. 31.
37. Ibid., pp. 36–9.
38. A short analysis of the influence of Nishiyama in disseminating Feder's writings in Japan appears in C. Hein, 2000, op. cit., pp. 146–7.
39. U. Nishiyama, 1968, op. cit., pp. 12–16, 24–5.
40. A. Flores Urushima, op. cit., p. 402.
41. U. Nishiyama, 'Risō toshiron' [A review of ideal cities], in Kyoto daigaku kōgakubu kenchikugaku kyōshitsu kenchikugaku kenkyūkai [Kyoto University, Faculty of Engineering, School of Architecture, Discipline of Architecture, Architectural Research Society], *Shinkenchiku no tenbō* [The prospect of a new architecture], Osaka: Naigai shuppansha, 1949.
42. U. Nishiyama, 1968, op. cit., pp. 577, 580.
43. The original text is U. Nishiyama, 1949, op. cit., pp. 139–40. The revised version is U. Nishiyama, 1968, op. cit., pp. 581–2. The quotation marks appear in both texts, of 1949 and 1968, without further explanation.
44. Ibid., p. 30.
45. M. Jay, 'Scopic regimes of modernity', in H. Foster (ed.) *Vision and Visuality,* Seattle: Bay Press, 1988, p. 3.
46. See the section 'Commodity, culture and alienation' and Muthesius's criticisms of public advertising in F. J. Schwartz, *The Werkbund: Design Theory and Mass Culture before the First World War,* New Haven: Yale University Press, 1996, pp. 44–60, 90–1.
47. H. Foster, op. cit., p. 100.

7

TOWNSCAPE AND SCENOGRAPHY

Conceptualizing and communicating the new urban landscape in British post-war planning

Peter J. Larkham and Keith D. Lilley

> Images are, in Kenneth Boulding's words, 'subjective knowledge'. To add to an understanding of the city-planning profession, close attention should be given to images held and purveyed by planners. The behavior both of planners and of their clients depends on images . . . [1]

Introduction

How people perceive the urban landscape is a major issue. How conceptions of the urban landscape are developed and employed, principally in communicating ideas in urban planning and design, is equally significant – and this includes communication within design teams that might be drawn from different professions, between professionals and elected representatives, and between all of these and the ultimate users, the wider public. Far more cities, districts and buildings exist on paper than have physical substance.[2] Multiple plans can be produced for the same location even within a short period of time, and in some periods – such as the post-Second World War reconstruction era – numerous plans for many places can be produced virtually simultaneously. This can cause confusion, doubt and delay in the minds of the public and professionals alike, as is suggested by the 1947 phrase 'a jungle of planning on paper'.[3]

A wide range of disciplines has seized upon this rich source material in various explorations of how cities are, or could have been, shaped at various periods.[4] These explorations have been both practical, in reviewing how and why certain proposals remained unbuilt, and conceptual, in determining the agencies and mechanisms of change, or elucidating the thought processes at work.[5] Inevitably, therefore, these paper plans – both the written text and the accompanying visual representations – are treated as 'text' amenable to a range of analyses.[6] Although this approach was not new in the period we examine, many earlier plans lacked such clear visual representations. The texts reviewed here are the series of 'reconstruction plans' produced during and immediately after the Second World War, in a complex cultural, social, economic and professional milieu that included unanimity of purpose against an enemy, overwhelming shortages, the emergence of the new professions of town planning and landscape design, and an expert-driven, almost exclusively male-dominated professional worldview. Our main focus is on the UK, although these concepts are equally applicable to the plans produced by UK-trained professionals in many other countries in the post-war period.

We use the term 'scenography' not to present architecture or urban form as a stage set (as, for example, in Foster's study of Schinkel)[7] nor in the sense of the 'picturesque', a version of which some have seen in the plans and products of this period.[8] Instead, this chapter explores aspects of the production and consumption of the imagery of the replanned city. It was *visual* imagery, the representation of the future urban landscape, which most powerfully and directly conveyed to the public the ideas of those who were involved in the process of reconstruction planning. It is important to recognize that those who played such a vital role in creating this imagery are often rendered almost invisible because their names do not appear on the front covers of the published plans. To speak of a particular consultant's plan for a particular place is thus virtually to silence the multitude of artists, perspectivists, photographers, modelmakers, cartographers and so forth whose work collectively underpinned that of the 'master planner' author.[9] That some commentators seem to have equated the illustrator with the designer adds to the confusion.[10]

A key concept gaining prominence at the time was recognition of the three-dimensional importance of representing replanned places – a 'scenographic' approach. In particular, in the terms of Reiner and Hindery, these are images *for* the city, suggesting an idea, artefact or process, or *in* the city, 'which by their very association with the specific give character and identity to place'.[11] Inevitably, there is the dichotomy of images of the real (what exists) and the ideal (what is proposed). We examine this particularly through the lens of the writing of Thomas Sharp.[12] Sharp is of particular interest owing to his prolific series of post-war reconstruction plans, his careful use of illustrative material and his noteworthy contribution to the emergence of the word and concept of 'townscape' in British planning. Clearly the latter has a history stretching back to Unwin and Sitte, but as reconceived by Sharp and his contemporaries it has achieved a much wider professional and lay circulation, and is still in common professional and lay use today.

The plans formulated so prolifically by Sharp and Abercrombie, and less prolifically by many others at this time, were, in reality, composites of various forms of text and visual imagery, each a 'plan' in its own right, which, when taken together, represented not one but a plurality of visions of the future city. Instead of taking these plans as evidence to help condemn the perceived 'misguided' vision of the master planners, as some have done, we wish to dig deeper into the planning process itself, examining how these 'advisory' reconstruction plans were produced in the difficult working conditions that prevailed in 1940s Britain, and their place in the emergence of 'townscape' as a new way of conceptualizing the urban landscape. The scenographic townscape representations in plans of this short period demonstrate a great engagement with visual communication of planning ideas, which then faded rapidly with the very different form for Development Plans that followed the 1947 Town and Country Planning Act.

The imagery of 1940s reconstruction plans: artists, aesthetics and the use of 'seductive' scenography

We are here concerned with the nature and range of the visual imagery used in the published reconstruction plans. We emphasize how the various forms of representation used in these plans reflected the importance of thinking about urban landscapes, especially in three dimensions. At the root of this was the use of seductive scenography, perspective views of 'what might be'.

Scenography is a form of visual representation. Vitruvius differentiated between *scenography,* 'the method of representing buildings perspectivally on a surface'; *ichnography,* 'the representation of the building in plan'; and *orthography,* 'the elevation' of a building.[13] Scenography thus has the power 'to show the sides as well as the façade of a building', whereas the other two forms of representation do not.[14] Scenography allows the artist to represent a townscape in three dimensions rather than just two. Most important, in this context, scenography allowed representation of the new (what is planned) in the context of the existing urban landscape. What scenography gave to the published reconstruction plans was visual immediacy, a sense of a future reality. Unlike two-dimensional representations such as maps, plans and elevations, the use of perspective drawings and photographs of models in reconstruction plans gave the viewer, especially the lay public, a more immediate and compelling impression of how things might be.

Most of the published plans made great use of scenography. The plans produced by both local authority departments and planning consultants were lavishly illustrated with perspective drawings. The more expensive published plans, such as Abercrombie's well-known plans for Plymouth, Kingston upon Hull and Edinburgh, generally made more use of colour reproductions of perspectives than did the lower-budget publications. Using a well-known commercial perspectivist could be extremely expensive, but even the consultants would quote several hundred pounds for undertaking the work themselves (£200 for nine perspectives for Worcester, for example)[15] (Figure 7.1). The local authorities' in-house publications usually contained less text and fewer maps, but sometimes more perspectives. To illustrate these 'popular' plans, they used not only perspective *drawings* but also *photographs* of display models – a cheaper method, perhaps, providing that the models had already been built; but for Hartlepool's plan exhibition, the model-building costs were £600, and the consultant's additional fee 'for the design of buildings and the direction of the models' was £210.[16] The widespread use of scenographic representation, especially artists' renderings, in these public planning documents requires some careful consideration, for the imagery used in the perspective drawings differed from one artist to another.

FIGURE 7.1 Seductive scenography: a perspective (original in colour) by Anthony Minoprio for the plan for Worcester by Minoprio and Spencely (1946), (courtesy of Worcester City Council).

It was often the case that the work of more than one artist appeared in published reconstruction plans. An example is the Royal Academy Planning Committee's *Road, Rail and River in London*. This was a successor to the perhaps better-known but critically mauled plan *London Replanned.*[17] The purpose of the plan, according to the preface, was 'to show that practical efficiency need not be divorced from orderly and tidy planning, and that such planning helps and encourages the design of fine buildings'. Like *London Replanned,* it contained many perspective drawings, some of which had been shown at the Royal Academy Summer Exhibition of 1944. One particularly striking page (Figure 7.2) shows two perspective drawings, both of proposed roundabouts, by P.D. Hepworth and A.C. Webb. Despite their common subject, the two drawings are very different. That by Hepworth is a low-level perspective, animated by buses and people in the streets, and brought to life by trees casting shadows and by building details. Although reproduced in black and white, Hepworth's drawing conveys colour and warmth. Webb's perspective, on the other hand, is an altogether more clinical and cold treatment, rather technical because of his use of labels such as 'No. 13 Radial'. The angle of vision is steeper and thus less attuned to human scale, while his streets are empty of traffic and people, and the buildings lack façades.

The juxtaposition of these two drawings presents a very striking contrast, perhaps intentionally. Although they reflect two extreme ways of visualizing townscape using scenography, in essence both drawings reflect a common purpose: to project future possibilities. Despite sharing this common aim, however, the two artists' 'plans' actually appear to compete with one another – for not only do they look so different, but they seem to be representing alternative versions of the future. This is particularly clear in *London Replanned,* as well as in other plans where perspective drawings were used in quantity. It presented a significant problem that had to be confronted: having so many different visions of the future might have prompted a reader to inquire which was the 'real' plan. Such questioning might destabilize the legitimacy of the plan itself, and thus even threaten the legitimacy of the planning process.

This raises the issue of expectations on the part of the plan readers. Perspective drawings had a lengthy history, particularly in architecture, but the planning context was different, and the scenographic perspectives had to be put into context – so readers had to be told that the drawings were there merely to illustrate ideas. In the 1945 Norwich plan it is stated several times that 'the new buildings shown are intended to indicate height and size only and are therefore made as non-committal in design as possible'.[18] Many plans extend this to more than architectural style. Hence, in a brochure published to accompany an exhibition of planning proposals for Swindon, the Mayor found it necessary to point out that 'we have deliberately . . . dealt with certain matters in outline only, we have avoided specifying in detail what development will take place many years hence, and we have fought shy of that sort of planning which can fairly be called daydreaming'.[19] Similarly, beneath three ground-level perspectives that appear in the *Plan for the New Coventry,* the caption informs us 'that at the moment it is the plan that is of primary importance, the details can follow later'.[20] This form of phrasing occurs frequently in plans that contain provocative perspective drawings or photographs of models.[21]

This reveals how far it was recognized at the time that perspective drawings were a powerful device, potentially dangerous to use but at the same time necessary if planning, and the large-scale and radical proposals suggested, was to be sold to a war-weary and rather sceptical public.[22] It might also account for the sometimes rather sketchy, ill-defined and dark-shaded views used in some plans, as well as the use of blank façades, evident always in A.C. Webb's drawings,

Drawn by P. D. Hepworth.

GENERAL VIEW, LOOKING EAST, OF ROUNDABOUT AT JUNCTION OF CITY ROAD AND OLD STREET.

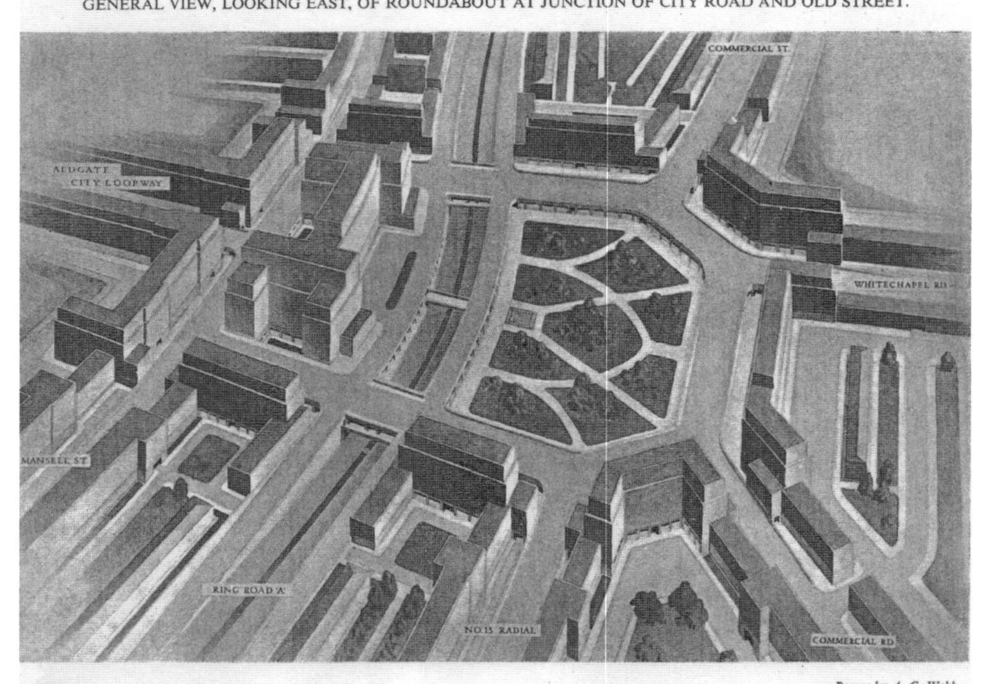

Drawn by A. C. Webb.

ISOMETRIC VIEW OF GARDINER'S CORNER ROUNDABOUT AT JUNCTION OF CITY LOOPWAY AND RING ROAD.

FIGURE 7.2 Two contrasting illustrations from the Royal Academy's *Road, Rail and River in London* (1944). Copyright Royal Academy of Arts, London.

which appear in many of Sharp's plans (Figure 7.3). Some perspectivists thus appear to have deliberately avoided committing themselves (and therefore their patron) by keeping drawings of proposed developments to outline form only. Equally, though, there were also those who drew scenes that made the drawings appear as life-like as possible, such as J.D.M. Harvey; and both types of images were to be found side by side in published plans.[23]

FIGURE 7.3 'A view of the redeveloped city from the west', aerial perspective of Exeter by A.C. Webb, for Sharp's 1946 plan (fold-out facing p. 98: original in colour).

The power of perspective drawings, recognized and understood by their creators, lay in their capacity to seduce the viewer. By its very nature scenography presents an illusion of reality: it renders concrete the artist's dream. Unlike other forms of visual representation used at the time, that is ichnography and orthography, scenography helped to give the published reconstruction plans a public appeal that they might otherwise have lacked. The value of the perspectives in this respect was, of course, exploited by the consultants and planning committees involved in producing the plans. However, while perspective drawings offered a compelling three-dimensional visualization of the townscape of the future city, they also presented difficulties: for perspectives could make plans look too real, and too many drawings (and, perhaps, different styles) could compromise and undermine the overall unity of a plan. Perspective drawings were thus a double-edged sword, capable of both knighting and beheading the 'master planner'.

The rise of 'townscape': concept and practice

The word 'townscape' has been noted in use as early as 1880 and 1889, and Bandini suggests that 'the term has several connotations: in architectural debates it is an analytical and prescriptive concept, and in urban geography it is a general term of reference'.[24] The term 'urban landscape' was used editorially in the *Architectural Review* in early 1944, the term 'townscape' in 1947 (by J.M. Richards et al.); and the journal's owner/editor used the term 'townscape' in a later contribution arguing for a picturesque informality of urban layout, contrasted to the then-dominant beaux-arts formal designs (H. de Cronin Hastings, writing as Ivor de Wolfe).[25]

Bandini discusses these as evidence of a 'picturesque' approach in English architectural polemic, in which 'townscape' became 'defined as a concept . . . the banner for this approach, providing it with a more concise method for analysis and design'.[26]

Thomas Sharp's early plans show great textual sensitivity to the qualities of individual places, particularly their physical structures and visual relationships.[27] However, he uses terms such as 'town landscape' in his Durham plan, the first in his significant series of reconstruction plans.[28] In his Exeter plan he discusses the town's 'genius' (i.e. *genius loci*) and its 'physical and architectural character' (the book's first section).[29] In conceptual terms, though, this appears to be identical with many later uses of 'townscape'. By his Oxford plan two years later, 'townscape' was the word used.[30] It is inconceivable that he did not know of the earlier *Architectural Review* articles, but here was a wider, more public introduction: 'by an analogy with an equivalent art practiced by the eighteenth-century Improver of land (we, after all, are Improvers of cities) it might be christened TOWNSCAPE' (his emphasis).[31] This plan and book were seen, in published and exhibition form, by tens of thousands of people.[32] Discussing urban and architectural form, and the inherent problems raised by the contrast between the physical experience of changing forms as one moves through a town and the static images that a book can provide, Sharp goes on to suggest that:

> . . . it would be worth considering every illustration in this book on its merit as a piece of Townscape. In the sense that the art of Townscape has always been practised without overt acknowledgement it is, of course, not a new art at all. Yet an art that is not consciously admitted remains incipient. Brought into consciousness, Townscape reveals possibilities which may conceivably make of it the representative art of this epoch.[33]

The clear formulation by Sharp (most clearly in the Oxford plan) and others of a 'townscape' term and concept came marginally before the most important series of articles in the *Architectural Review* (spearheaded by Gordon Cullen, the journal's Art Editor in 1949–56 and himself the illustrator of some reconstruction proposals) that publicized and hardened the concept of 'townscape' in the late 1940s and 1950s. It became widely accepted, with the *Architectural Review*'s 'townscape manifesto' being reprinted with favourable comment in the *Journal of the Town Planning Institute* in 1954.[34] Cullen developed the term as a visual analytical tool, using sketches and informative captions, and the concept of 'serial vision' (prefigured perhaps by Sharp's comments in the Oxford plan) first in these short illustrated reports, then in a full-length book.[35] Cullen's sketches, and his serial vision perambulations, are clearly an application of scenography, but he usually used it analytically to identify qualities of places (present and future, good and bad) rather than mere static representations of 'what might be'. Yet even in his oft-cited and still available *Townscape,* there is but one indexed use of the term, and that is particularly vague: 'If I were asked to define townscape I would say that one building is architecture but two buildings is townscape.'[36]

By 1949, Sharp was using 'townscape' in public lectures and speeches. His Presidential Address to the Institute of Landscape Architects used the word frequently and defined it as 'urban scenery considered as the combination, the integration, of the open and built-over parts of the town, and within that combination the effective interplay of the various elements of both parts'.[37] Four years later the term and concept were sufficiently embedded in contemporary planning that they formed a major discussion theme at the Town and Country Planning Summer School. Here, 'the fundamental elements of townscape are the various spaces in the town; the proportions of its streets, squares and places as seen by the "man in the street"' – however, 'the subject of townscape is one that is not well understood in this country and there is a great need for further study'.[38]

It took Sharp nearly two decades from the date of his reconstruction plans to refine and, most significantly, to publish more widely his own conceptions of townscape as 'a way of looking' in *Town and Townscape*.[39] This approach is important since it focuses our attention on the *process* rather than solely on the scenic *product*. Only in a footnote in this book does Sharp explain:

> The word 'townscape' has come into use of recent years among town-planners and architects. But it has generally been used either to denote a single street-scene or the collection of elements that constitute it – elements varying in size from large objects like buildings to small ones of every kind . . . This imprecise use of the term is confusing. It would be better to stick to the term *street-scene* for the single scene, keeping *townscape* to mean the wider interconnecting ones that constitute town scenery.[40]

For him, townscape was not the analytical tool of Cullen's drawings, the 'philosophy of urban design' as Whistler and Reed later termed it.[41] Sharp's examples show that he uses 'townscape' still in the kinetic sense, but incorporating micro- and macro-scale observation and sensitivity to local character, including knowledge of history, topography and contemporary planning issues including traffic. It is these very qualities that are praised in many reviews of his 1940s plans.

Visualizing townscape: planning in three dimensions

On close examination of these 'popular' advisory reconstruction plans, we find a preferred use of three-dimensional visual imagery, particularly perspective sketches and photographs of models, rather than two-dimensional images, such as plans and maps.[42] This emphasis on three-dimensional planning should be no surprise, for it was an approach that professional planners and architects in Britain were increasingly advocating during the war. In two articles in the *Architectural Review*, for example, the Tatton-Browns set out their ideas on street-level planning.[43] In April 1945 the Modelling Unit of the Ministry of Town and Country Planning held a display of photographs of town models, which showed:

> how the camera can pick out and throw into relief separately points of special interest; give added realism by excluding disproportionate or irrelevant surroundings; record the incidence of light and shadow; and, by reproduction and display, multiply the audience to which the model can appeal.[44]

Donald Gibson, principal designer for the devastated Coventry, noted that much contemporary planning dealt only with two dimensions, but 'whether or not our towns are to form a beautiful and satisfying environment will depend very largely upon planning in the third dimension'.[45] He advocated the more frequent use of models, which were 'very useful both for the public, who generally cannot understand plans very easily, and also for the developer and for the designer himself'.[46]

In his influential wartime book *Town Planning*, Sharp had already derided the 'paper patterns' and 'silly symmetry' of those planners' plans that 'could only be appreciated from above', from the air (thus echoing the distant words of Raymond Unwin some 40 years before).[47] The beaux-arts symmetry and pattern of the Paton Watson and Abercrombie plan for bombed Plymouth, although published three years later, appears to fit this category (Figure 7.4).[48] In his Durham plan, Sharp noted that the city's topography means that many areas and buildings are naturally viewed from above, and therefore need to be conceived in three dimensions.[49] But other, flatter cities received the same treatment. In fact, Sharp's concerns were for an approach to planning that valued towns as a whole, including aerial views and skylines as means of conceptualizing the functions as well as the aesthetics of the place.

Sharp's textual sympathies for qualities of landscape, most particularly in the irregularly planned mediaeval cathedral towns rather than the more regular mediaeval and Georgian towns in which he also worked,[50] led to the development of his concept of 'kinetic townscape' and explicit recognition of the 'mutability' of the urban landscape.

> If one can agree to discard the old conception which sees the urban scene as a series of stills and learn to regard it as a *mobile* – a moving visual target – in which the constituent motifs constantly reassemble under their own power . . . one begins to get an idea of the possibilities of the art of civic design.[51]

For his Oxford plan, Sharp set out in detail an example of 'kinetic townscape', the basic principle of which was the idea that the urban landscape had to be appreciated from the street level, particularly from the pedestrian's perspective.[52] This links well with Gibson's view that 'an aeroplane view of a town is important, but it is the view from the street which should be the over-riding consideration'.[53] Only then was it possible to really appreciate the contrasts, or 'foils', which, in Sharp's view, were the essence of visual quality in the townscape.[54] This struck a chord with one reviewer:

> What first excited my interest in Mr Sharp's book was a series of photographs with the caption 'Moving Target'. They show that one's view of a group of buildings is kinetic, since the buildings are in continual movement in relation to each other and to the observer as he walks down a street.[55]

Such was the concern for thinking about the townscape in three dimensions in 1940s planning. 'Kinetic townscape' prefigures some of the ideas implicit in Logie's general text of the mid-1950s on the 'urban scene' and the better-known concept of 'serial vision' developed by Cullen from 1949.[56]

Two apparent contradictions emerge here. Both are embedded in, and constitutive of, the traditional discourse of planning history and the perception of urban landscapes. The first is the image of the master planner as a technocratic tyrant, imposing his 'view from above'. This, common as a phrase and a critique, can be interpreted both as the style of top-down technocratic planning, and as referring to the representation of planning by maps, which became increasingly common and stylized after the 1947 Town and Country Planning Act and its associated prescriptive guidance on the content and form of Development Plans.[57] Thus British planning moved away from the scenography/landscape style of representation towards the professional mystique of the map and data tabulation, to the extent that 'by the 1960s, post-war reconstruction planning had become no more than totalitarian, authoritative and statistical'.[58] British planning moved away from 'paper cities' and produced dry statistical documents much less comprehensible to a lay readership. This appears to be an altogether different sort of approach

FIGURE 7.4 The iconic plan for central Plymouth (Paton Watson and Abercrombie, 1943: original in colour). Is this an example of Sharp's 'silly symmetry': 'it can only be appreciated on paper or from an aeroplane. And most people spend little of their time looking at maps or hovering . . . in aircraft'? (Sharp, 1940, *Town Planning*, p. 105).

to conceptualizing and representing planning from, for example, that of Sharp, with his clearly expressed ideas about street-level, three-dimensional planning. It is likely that academic and popular planning histories have placed too much emphasis on the idea that the so-called master planners of the 1940s represented a modernist *zeitgeist* – and, by the 1970s, were judged to have been out of touch and ultimately doomed. As Gold demonstrates, there is a common and misguided view of modernism that does not recognize its subtleties and variety.[59] The second contradiction, again to do with three-dimensional planning, is perhaps more interesting. Sharp's reconstruction plans for historic English cathedral cities and market towns show it very clearly. Here, the extensive use of low-level aerial perspectives – indeed, some drawn by the expert perspectivist A.C. Webb from aerial photographs to Sharp's precise instructions (cf. Figure 7.3) – to visualize the form of new townscapes appears to be somewhat at odds with Sharp's own arguments against 'paper patterns' and 'silly symmetries'. In this respect, what can be read in the planner's text and what can be seen in the *imagery* of the plan seem to be at odds. In order to make sense of this, we have to recognize that embedded within each 'plan' there are, in fact, multiple plans. Even Sharp, who had a very small staff and drew many of his own diagrams, used several artists, such as Webb and S.R. Badmin, to produce perspectives,[60] together with modelmakers and photographers; and the politicians and professionals of the commissioning organization obviously also had some input. Each plan also had multiple functions, including expressing complex concepts clearly to a broad range of readerships.

Such complexities and contradictions in the popular, published reconstruction plans of the 1940s lead us to look 'behind' the plan; to peer into the process of plan production in general, and look in particular at the ways in which visual imagery was used. This chapter does so by examining scenographic representation in the nature of the visual imagery used in British reconstruction plans in the 1940s, by addressing what processes were involved in image production and by exploring the rise of 'townscape' and several uses of it in professional discourse. We demonstrate, first, that the so-called master plan needs to be thought of more critically, in terms of its content, its production and its use as a communication tool; and, second, that the idea that 1940s planning was all about 'master planners' imposing their 'view from above' requires some careful reconsideration and revision. This has led to some reflection on the changing conceptions of the urban landscape and planning approaches to it in the early post-war era.

Two things become apparent about the inter-relationships between clients, consultants, artists and modelmakers. First, any reconstruction plan represents a series of 'plans', including the concept of the client, the text of the author (consultant, local authority officer or, more rarely, member of the public or voluntary organization) and the conceptions of the illustrators. Second, the illustrations developed the concept of scenography as a means of representing new – sometimes radically so – urban forms to clients and the general public in an attractive and convincing way. Although a technocentric, top-down approach to planning was used at this time, clearly there was a strong view that the public had to be both informed and convinced about the merits of what was being proposed.[61] Other forms of representation, including cartographic and architectural elevation, were used, but more frequently to make technical/professional points, rather than as a means of convincing the public in brochures and exhibitions. Scenography, planning in three dimensions, captured the public imagination, for it gave the viewing public the 'eyes' of the architects and planners – they, too, could look down from above, as well as using imagined eye-level perspectives, and see for themselves the townscape of the city of tomorrow. Although this was not wholly new, this short but critical period of intense planning activity and public engagement led to views such as those expressed in one professional journal, as early as 1941, that:

> visions of reconstruction, of new and better worlds, of shining open cities with parks and spaces and glorious avenues offer irresistible temptation to the dreamers, the orators,

and to public men in general . . . What is required is an outline, even if sketchy, of these future things which are actual possibilities, presented in such a form that the average intelligence can grasp their impact.[62]

Some have argued that the scenography/townscape tradition died out in British planning,[63] and certainly the tremendous change in format and direction from the reconstruction-era 'outline plans' to the technocratic 1947 Act's Development Plans would support this. Post-1947 Act Development Plans focused on mapping and land-use issues: the third dimension in planning was relegated to piecemeal consideration through development control mechanisms. The rise of urban design considerations in recent years provides some encouragement through the enduring popularity of Cullen's *Townscape* and the scenographic illustrations in many contemporary urban design proposals. In some respects also, Sharp's fame endures: the contemporary historian Tristram Hunt described him in *The Observer* as a 'vogueish post-war planner';[64] yet he seems now almost forgotten within his profession despite his significant contribution to developing the practice and communication of planning as discussed here.

Acknowledgement

This research has been supported by the British Academy and the Leverhulme Trust. We are grateful to colleagues and conference participants who have made suggestions on various versions of this text.

Notes

1. Reiner, T.A. and Hindery, M.A. (1984) 'City planning: images of the ideal and the existing city', in Rodwin, L. and Hollister, R.M. (eds) *Cities of the Mind: Images and Themes of the City in the Social Sciences,* New York: Plenum, p. 133; citing Boulding, K. (1956) *The Image,* Ann Arbor: University of Michigan Press.
2. Söderström, O. (1996) 'Paper cities: visual thinking in urban planning', *Ecumene,* 3: 249–81.
3. 'Delay in City of London rebuilding' (1947) *Estates Gazette,* 4 October, p. 253.
4. For example, in geography, Whitehand, J.W.R. (1992) *The Making of the Urban Landscape,* Oxford: Blackwell; in planning and architectural history, Barker, F. and Hyde, R. (1982) *London as it Might Have Been,* London: Murray.
5. An example of the latter is Söderström, 'Paper cities'.
6. Cf. Barnes, T. and Duncan, J. (eds) (1992) *Writing Worlds: Discourse, Text and Metaphor in the Representation of Landscape,* London: Routledge.
7. Foster, K. (1994) 'Only things that stir the imagination: Schinkel as scenographer', in Zukowsky, J. (ed.) *Karl Friedrich Schinkel: the Drama of Architecture,* Chicago: The Art Institute of Chicago.
8. See Horton, I. (2000) 'Pervasion of the picturesque', in Hughes, J. and Sadler, S. (eds) *Non-Plan,* Oxford: Architectural Press. Contemporary uses of the picturesque include 'The English planning tradition in the city' (1945) *Architectural Review,* June: 165–76; the art historian Kenneth Clark (quoted in Woodward, C., 2001, *In Ruins,* London: Vintage, p. 212); and the architect Ralph Tubbs (1942, *Living in Cities,* London: Penguin). Other aspects of this debate are covered elsewhere in this volume.
9. Even if not explicitly acknowledged in print, many authors were privately grateful for the assistance of illustrators: a manuscript note in Peter Larkham's copy of the Aberdeen plan (Chapman, W.D. and Riley, C.F., 1952, *Granite City: a Plan for Aberdeen,* London: Batsford), signed by W. Dobson Chapman, reads 'To Peter G. Elphick with compliments and many thanks for your assistance on the model drawings etc. which helped to make the work possible' (dated 23 December 1952).
10. See Barker and Hyde, *London as it Might Have Been,* p. 179.
11. Reiner and Hindery, *City Planning,* p. 135.
12. Stansfield, K. (1981) 'Thomas Sharp 1901–1978', in Cherry, G.E. (ed.) *Pioneers in British Town Planning,* London: Architectural Press; Cherry, G.E. (1983) 'Thomas Sharp: the man who dared to be

different', Sharp Memorial Lecture, University of Newcastle upon Tyne (copy of Cherry's text in the authors' possession); see also Pendlebury, this volume.
13. Cf. Wood, C.S. (1997) 'Introduction', in Panofsky, E. *Perspective as Symbolic Form,* New York: Zone Books, p. 97.
14. Ibid.
15. Reconstruction and Development Committee Minutes, 13 January 1946, Worcester County Record Office.
16. *Northern Daily Mail* (1947) 7 April: item on exhibition for Lock's plan for Hartlepool.
17. Royal Academy Planning Committee (1942) *London Replanned,* London: Country Life; Royal Academy Planning Committee (1944) *Road, Rail and River in London,* London: Country Life.
18. James, C.H., Pierce, S.R. and Rowley, H.C. (1945) *City of Norwich Plan,* Norwich: City of Norwich Corporation, for example p. 51.
19. Preface to Davidge, W.R. (1945) *Planning for Swindon,* Swindon: Swindon Borough Council.
20. Gibson, D. (1945) *Plan for the New Coventry,* pamphlet circulated by Coventry City Council, reprinted from *Architect and Building News,* 1945.
21. Clearly some plans were more innovative in their use of photography, with Sharp's as published by the Architectural Press being in the forefront. Yet the scenographic visions of the future produced by perspectivists and models were crucial to a wider understanding of these proposals.
22. The scepticism is shown, for example, by the lack of positive public reaction to Wolverhampton's plan (Larkham, P.J., 2002, 'Reconstructing the industrial town: wartime Wolverhampton', *Urban History,* 29: 388–409) and by the 1944 comment from Manchester that '[t]here's a curious lack of interest in the city plan. People are distracted with other things and are glad to think of it only as "tentative"' (James, B., 1944, 'Letter to the Editor', *Manchester Guardian,* 15 August).
23. The artists are discussed further by Stamp, G. (1982) *The Great Perspectivists,* London: Trefoil; and Larkham, P.J. (2007) 'Selling the future city: images in UK post-war reconstruction plans', in Whyte, I.B. (ed.) *Man-Made Future: Planning, Education and Design in Mid-Twentieth-Century Britain,* London: Routledge.
24. Bandini, M. (1992) 'Some architectural approaches to urban form', in Whitehand, J.W.R. and Larkham, P.J. (eds) *Urban Landscapes: International Perspectives,* London: Routledge, p. 166, note 1.
25. 'Exterior furnishing or Sharawaggi: the art of making urban landscape' (1944) *Architectural Review,* 95: 125–9 (apparently authored by H. de Cronin Hastings); The Editors (1947) 'The second half century', *Architectural Review,* 101, special supplement; de Wolfe, I. (pseudonym of Hastings) (1949) 'Townscape: a plea for an English philosophy founded on the fine rock of Sir Uvedale Price', *Architectural Review,* 106: 355–62. See also Erten, E. (2009) 'Thomas Sharp's collaboration with H. de C. Hastings: the foundation of townscape as urban design pedagogy', *Planning Perspectives,* 24: 29–49, and this volume.
26. Bandini, 'Some architectural approaches', pp. 140–1; see also Horton, 'Pervasion of the picturesque'.
27. Pendlebury, J. (2009) 'Thomas Sharp and the modern townscape', *Planning Perspectives,* 24: 3–27.
28. Sharp, T. (1944) *Cathedral City: a Plan for Durham,* London: Architectural Press, p. 85.
29. Sharp, T. (1946) *Exeter Phoenix,* London: Architectural Press. The Exeter-based exhibition of this plan alone was visited by about one-third of the city's population (*Western Morning News,* 1946, 15 January).
30. Sharp, T. (1948) *Oxford Replanned,* London: Architectural Press, p. 34.
31. Ibid., p. 36.
32. The exhibition is discussed in *Architect and Building News* (1948), 193, pp. 235, 284–8, 389; the plan and/or exhibition are reviewed in at least ten contemporary professional journals and wider-readership magazines.
33. Sharp, *Oxford,* p. 36.
34. The introductory sentence to this reprint quotes the *Architectural Review* of 1953 thus: 'It is remarkable how little attention is paid to Townscape, even by those who have some professional interest in it, and it is not at all surprising, therefore, that the general public's interest should be negligible' ('Townscape manifesto', 1954, *Journal of the Town Planning Institute,* 30: 262).
35. For example Cullen, G. (1949) 'Townscape casebook', *Architectural Review,* 95: 125–9; Cullen, G. (1961) *Townscape,* London: Architectural Press.
36. Cullen, *Townscape,* p. 133.
37. Quoted in 'Landscape and townscape' (extracts from Sharp's Presidential Address to the Institute of Landscape Architects) (1949), *Estates Gazette,* 19 November, p. 449.
38. Wilson, L.H. (1953) 'Townscape', *Town and Country Summer School 1953, Report of Proceedings,* London: Town Planning Institute, pp. 103, 105.

39. Sharp, T. (1968) *Town and Townscape,* London: Murray, chap. 3.
40. Ibid., p. 40.
41. Whistler, W.M. and Reed, D. (1994) 'Townscape as a philosophy of urban design', *Urban Design Quarterly,* 52: 16–19 (a partial reprint from a 1977 bibliography for the Council of Planning Libraries Exchange).
42. The varying importance of these different forms of representation is well shown by Perkins, C. and Dodge, M. (2012) 'Mapping the imagined future: the roles of visual representation in the 1945 City of Manchester Plan', *Bulletin of the John Rylands University Library,* 89: 247–276.
43. Tatton-Brown, A. and Tatton-Brown, W. (1941) 'Three-dimensional town planning', *Architectural Review,* 89: 17–20; (1942), 90: 82–8.
44. 'Photography and models' (1945), *Estates Gazette,* 14 April, p. 292.
45. Gibson, D. (1946) 'Letter to the Editor', *Architects' Journal,* 3 October, p. 240.
46. Gibson, D. (1947) 'The third dimension in town planning', paper presented at the Town and Country Planning Summer School, Reading, excerpted in *Architect and Building News,* 1 August: 97–100.
47. Sharp, T. (1940) *Town Planning,* Harmondsworth, Middlesex: Pelican, p. 88. This is said to have sold a quarter of a million copies, an enormous number for wartime: the planning historian Gordon Cherry wrote that it was perhaps 'the planning bestseller of our time' (Cherry, G.E., 1974, *The Evolution of British Town Planning,* London: Leonard Hill, p. 130).
48. Paton Watson, J. and Abercrombie, P. (1943) *A Plan for Plymouth,* Plymouth: Underhill.
49. Sharp, *Durham.*
50. Lilley, K.D. (1999) 'Modern visions of the medieval city: competing conceptions of urbanism in European civic design', *Environment and Planning B,* 26: 427–46.
51. Sharp, *Oxford,* p. 34.
52. Ibid.
53. Gibson, 'The third dimension in town planning', p. 99.
54. Sharp, *Oxford,* pp. 32–3.
55. van Heyningen, W.E. (1949) 'Review of T. Sharp, *Oxford Replanned,* London: Architectural Press, 1948', *Town Planning Review,* 20, p. 89.
56. Logie, G. (1954) *The Urban Scene,* London: Faber; Cullen, 'Townscape casebook'; the idea was expanded in Cullen, *Townscape.*
57. See, for example, Ministry of Town and Country Planning (1948) *The Town and Country Planning (Development Plans) Regulations 1948,* SI 1948 no. 1767, London: HMSO.
58. Bartram, R. and Shobrook, S. (2001) 'Body beautiful: medical aesthetics and the reconstruction of urban Britain in the 1940s', *Landscape Research,* 26: 119–35, at p. 132.
59. Gold, J.R. (1997) *The Experience of Modernism,* London: Spon; Gold, J.R. (2007) *The Practice of Modernism,* London: Routledge.
60. Larkham, 'Selling the future city'.
61. Lilley, K.D. and Larkham, P.J. (2007) *Exhibiting Planning: Communication and Public Involvement in British Post-War Reconstruction,* Working Paper 4, Birmingham: Faculty of Law, Humanities, Development and Society, Birmingham City University.
62. *Architect and Building News* (1941) Editorial comment, 6 June, p. 131.
63. Maxwell, I. (1976) 'An eye for an I: the failure of the townscape tradition', *Architectural Design,* 46: 534–7.
64. Hunt, T. (2002) 'We must make our cities slicker', *The Observer,* 27 October, available online at <*http://society.guardian.co.uk/urbandesign/comment*> (accessed 3 January 2010).

PART III

Townscapes in practice

8

MAKING THE MODERN TOWNSCAPE

The reconstruction plans of Thomas Sharp

John Pendlebury

Introduction

In the mid-twentieth century British town planning was developing from its origins in civic design, garden cities and regulation of basic standards to become a more all-encompassing project for refashioning town and country on rational, modern lines, integrating issues of design with the emerging social sciences. While garden city principles were dominant in the emergent planning profession, quite different ideas were being inspired by Le Corbusier's radical ideas about future urban form.

Amongst all this, Thomas Sharp, a prominent figure in the British planning profession at this period, swam counter to some of the currents of fashion. Virulently anti-garden city and anti-suburb, and more sympathetic to – but ultimately dismissive of – Corbusian-type abstract models, he promoted a practical urbanism which, while drawing strength from Enlightenment models, was self-consciously modern in character. Visual planning was central to Sharp's approach, and one of Sharp's fundamental urban building blocks was the street, recovered from the debasements, as he saw it, of the nineteenth century. Sharp was both a polemical writer and, for a fairly short period, a prodigiously prolific producer of plans. This chapter will focus on the so-called reconstruction plans produced by Sharp between 1943 and 1950; part of an extraordinary body of such work.[1] Sharp was particularly known for his work on historic cities, with his plans for Durham, Exeter and Oxford the best-known trio. In this period Sharp's consultancy work also included a hypothetical new town for the Bournville Trust, the first masterplan for the new town of Crawley and the design of new villages in the remote Northumberland countryside for the Forestry Commission.[2]

The chapter is based on analysis of the plans themselves and on Sharp's private papers, held as a Special Collection at Newcastle University. The principal focus is on his developing ideas of kinetic townscape and composition. Central within this are some of his planning ideas, such as the role of the street as the primary urban building block and his approach to accommodating urban traffic.

Formative principles

Sharp announced himself to the wider planning world through his book *Town and Countryside,* published in 1932.[3] It is an important book in understanding Sharp and his views, for while he went on to refine and develop his ideas and perhaps expressed them better in subsequent texts, *Town and Countryside* set out core values which would be sustained throughout his career. Thus, common principles were consistently reiterated through his subsequent 'general' planning books, *English Panorama, Town Planning* and the substantially revised second edition of *English Panorama* (though this latter had some interesting shifts and changes of nuance from the first edition).[4]

Sharp set out his stall early in *Town and Countryside.* In part his argument stemmed from the widely held concern of the period over the perceived desecration of the countryside, as motor traffic allowed the ugliness hitherto largely associated with and confined to the industrial town to spill out into rural areas. In this respect he was following a path beginning to be well developed by others, such as Clough Williams-Ellis in *England and the Octopus,*[5] and represented, for example, by the formation of the Council for the Preservation of Rural England in 1926.[6] However, for Sharp, the problems of the future of the countryside were inextricably linked with the future of the town. Urban areas had lost urbanity, according to Sharp, because of Victorian industrialism and capitalism, but also because of the planning response of garden cities and their suburban progeny. Controversially for the time, he directly savaged Howard's *Garden Cities of Tomorrow.* The essence of his argument was that the correct response to the horrors of the Victorian city, rather than abandoning it, should rather be to suggest how it could be improved. He lambasted the concept of Howard's idea of the marriage of town and country as 'Town-Country' as being 'a hermaphrodite; sterile, imbecile, a monster; abhorrent and loathsome to the Nature which he worships'.[7] He lamented how the low-density ethos of the garden city had been encapsulated in planning legislation and in the profession. Howard was attacked perhaps to an even greater degree in *English Panorama,* while *Town Planning* continued the withering critique of garden cities (now labelled 'Neither-Town-Nor-Country') and suburbanization.

Sharp's alternative to the garden city was the rehabilitation of the idea of the town. Unlike some commentators, Sharp did not consider poor town building as an intrinsically English failing.[8] Indeed, he considered there to be a distinguished post-Enlightenment, pre-Victorian English history in this regard. This history was related to practical democratic utility rather than authoritarian show (which he contrasted with examples from mainland Europe). London squares exemplified this. Similarly, provincial towns in England (and some in Scotland) were compared favourably to their European counterparts, as pleasant towns for citizens (although, as Sharp made clear, historically not all citizens benefited). The results in both town and country were held by Sharp to combine utility and beauty. He was part of a vanguard, following Christopher Hussey, rediscovering and celebrating 'the picturesque' and mobilizing the concept as part of a contemporary approach to planning and design.[9]

However, though Sharp's appreciation of urbanity and urban form was rooted in the past, he was at heart a modernist. Rather than attempting to imitate the past, it should be used as a source of inspiration, not least in the boldness displayed by earlier generations in their urban interventions. Sharp's published views on Le Corbusier were somewhat equivocal. In *Town and Countryside* he referred to 'the much-discussed frenzied theatricality that Le Corbusier has entitled "The City of Tomorrow"'.[10] In his discussion of theorists in *Town Planning* he displayed some sympathy for Corbusian ideas but considered them impractical. He was not especially adverse to high-rise flats at this point and concluded the appropriate residential mix would be a combination of flats and houses. Indeed, his 1943 scheme for the Bournville Village Trust included one point block, prefiguring Frederick Gibberd's design for Harlow,[11] although later he wrote witheringly about the fashion for building tall.[12] However, from the time of *English Panorama* Sharp had pronounced, in direct contradiction to Corbusian thinking, that the key urban building block should be the street. His advocacy for the street, inspired by Georgian precedents, was also used to attack garden city-derived, semi-detached, hip-roofed, 'open' development – the prevalent form of new private and public housing at that time.

Having set out the case for why the later nineteenth-century street was a debased architectural form, *Town Planning* analysed its author's preference for the terraced street, as a form of urban design that provided the best picturesque architectural composition. As he had in *Town and Countryside* earlier, Sharp cited Trystan Edwards' *Good and Bad Manners in Architecture* as a key influence in this regard; in particular, he underlined its emphasis on doorways as an expression of the individual house.[13] Sharp outlined compositional principles for terraces which might avoid the monotony of the nineteenth-century street. For Sharp, each individual street should be regarded as an architectural composition, and a town should be a continuous series of contrasting compositions. Within an urban hierarchy, some principal streets might be quite long and given modest monumentality, but most would be short and might be culs-de-sac. Above all, the key was held to be variety. This consideration of the street was part of a wider approach to urban design that Sharp was formulating, and which he came to call 'townscape'; a term which of course was also used by a group of writers at the *Architectural Review,* ultimately leading to Gordon Cullen's seminal text.[14] I return to Sharp's connections with the *Architectural Review* below.

Similarly, the partially completed draft manual *Civic Design,* prepared for the Ministry of Town and Country Planning, opened a discussion on the street with the bald statement that 'the axiom [is] that the street is the urban unit of design'.[15] At the core of this was the importance of good neighbourliness between buildings: 'each street must be judged, and should so be designed, as a large-scale finite composition, a single urban picture'.[16] Monotony could be avoided by keeping streets relatively short and maintaining variety between streets. Formal architectural 'stops' were not deemed necessary, but there was considered to be a danger of anti-climax if compositional issues were not fully considered. The manual further considered issues relating to domestic streets and commercial streets. The latter included a blast against 'chain-store architecture', the practice of using standardized designs for particular companies, considering it 'a most deplorable abrogation of civic responsibility'.[17] 'Each town differs from every other town, every site from every other site and every individual problem of design requires its own individual solution.'[18]

Thus, through Sharp's writings we can see clear principles relating to urban organization that critiqued prevailing conventions (garden cities and Corbusian modernism) and proposed alternative forms of urban organization based around the primary building block of the street and an understanding of the individual characteristics of place. These views on urban form were underpinned by arguments for mixed use and against planning as means of achieving social segregation.[19] Much of what Sharp argued for would subsequently become conventional wisdom within the emergent field of urban design.

The reconstruction plans

From the late 1930s, apart from a period of secondment to the fledgling Ministry of Town and Country Planning, Sharp had been teaching planning at the School of Architecture in Newcastle upon Tyne, at that time part of the University of Durham. In 1945, seeing no prospect of academic advancement, Sharp resigned his teaching post to work as a planning consultant. He had in fact already been working on his first significant commission, his reconstruction plan[20] for the city of Durham. The Durham commission was followed by his engagement to prepare plans for Todmorden (a small industrial town) and for the historic but bomb-damaged Exeter. The following years produced jobs for other historic towns. The most significant of these was Oxford, but historic towns became something of a speciality – other plans were for King's Lynn, Taunton, Salisbury, Chichester and St Andrews. He also produced plans for Stockport and Minehead, towns not obviously of historic significance, and neighbourhood layouts for parts of Kensington and Hemel Hempstead.

In terms of sheer quantity of plans, Sharp can be regarded as the most prolific of the reconstruction planners. At the same time, he did not run a large office (unlike, say, Patrick Abercrombie). He stated that he never employed more than three assistants at any one time in his office and that he wrote all the text of the plans himself;[21] he also seems to have drawn many of his own maps. Perhaps the most obvious external contribution to his plans was the engagement of architectural illustrators, such as A. C. Webb, to produce, under careful instruction, visualizations of plan proposals.

Sharp himself placed most weight on the plans for Durham, Exeter and Oxford in his unpublished autobiography.[22] These represent three of the four Sharp plans which were produced with the highest production values, being published (along with the plan for Salisbury) by the Architectural Press – the other plans which were published were more home-spun, local publications. It is also in these three plans that we can see Sharp investing his own time and effort most passionately, often beyond the limits of what would have been commercially sensible, and seeking to promote wider manifestos of planning. In particular, we can see the evolution of his developing ideas of place, which become partly mobilized around the idea of townscape.

As has been remarked, Durham was Sharp's first professional commission. He was appointed to the job in December 1943, and the report was subsequently published as *Cathedral City*.[23] His was not an uncontroversial or unopposed appointment, probably because he had written critically of the County Council's proposals for the city as part of his contribution to the polemical text *Britain and the Beast,* and in particular about proposals for an elevated relief road.[24] Thus in Durham (and elsewhere) Sharp regarded the Second World War as a useful pause to enable a more considered approach to planning to be taken, something he is explicit about right at the start of the preamble to the plan. His appreciation of Durham was largely based on its visual qualities. The Cathedral, and to some degree the Castle, was central to this, though there was also a wider appreciation of the 'picturesque' and 'medieval' flavour of the city,[25] especially in terms of the roofscape and of the foil that domestic-scaled building gave to the major monuments. Sharp also stressed the historic importance of the Cathedral as part of emphasizing the significance of Durham, in order to resist local pressure for developments such as a massive power station proposal. Alongside a romantic engraving of the city he argued that the setting of Durham Cathedral was of such importance that 'the question of its mutilation becomes a matter of moment not merely to Durham or Britain but to Christendom'.[26]

Proposals for preservation were naturally focused on the peninsula containing the Cathedral, the Castle and heart of the University, and the commercial centre. The setting of the Cathedral and Castle was given extensive discussion. The setting was said to be formed by five elements, which included three that were more obvious: the river banks, the College (to the south of the Cathedral) and Palace Green (between the Cathedral and Castle). The two less-obvious elements were the Bailey (the street that runs the length of the peninsula) and Owengate, the short street that links the Bailey with Palace Green (see figure 8.1). The domestic Georgian character of the Bailey was regarded as a valuable foil to the 'massive dignity of the Cathedral'.[27] Ascending from the Bailey, Owengate

> climbs steeply up to Palace Green, with a glimpse of the Cathedral at its head. Then, at the top of the rise, at the head of the curve, the confined view having thus far excited one's feelings of mystery and expectation, the street suddenly opens out into Palace Green, broad, spacious, elevated, with a wide expanse of sky: and there, suddenly, dramatically, the whole fine length of the Cathedral is displayed to the immediate view. It is as exciting a piece of town planning as occurs anywhere in the kingdom.[28]

Sharp also saw merit in the wider peninsula. For example, the Market Place was held to have a sturdy character worthy of maintaining, although no individual buildings were considered to have any particular distinction. Beyond the peninsula he saw the need for extensive rebuilding, whilst acknowledging that some of the buildings to be cleared had architectural merit. Substantial construction was proposed, especially new buildings for public uses. New buildings were to be both identifiably modern and sensitive to place-character, but were not prescribed in detailed terms in the plan: this was to be a later task for individual architects. Sharp gave general guidance only; for example in the case of Durham he regarded roofscape and the use of pitched roofs as being vital (figure 8.2). Finally, a key part of the proposals was his alternative to the County Council inner relief road. He produced an extensive critique of this proposed road, which would have run on an elevated viaduct, in terms of its impact on the character of Durham, saying of his low-level alternative that 'it will *belong*'.[29]

Thus in *Cathedral City* we can see ideas of townscape beginning to develop, something particularly evident in the Owengate citation above, where an appreciation of the particular qualities of place and the way they are experienced in pedestrian movement through space is demonstrated. We also see him advocating a distinct and different approach to the problem of accommodating traffic in historic, tightly knit towns. This is the first of his proposals for a 'substitute road', an idea I return to below.

While working on Durham Sharp was engaged to prepare a plan for Exeter, published as *Exeter Phoenix*.[30] Exeter was the only one of the historic cities that Sharp worked on to receive major war-time damage; Sharp estimated that the city had lost something like half of its buildings of architectural merit through bombing. The report discussed at length those buildings destroyed, especially the Georgian buildings, such as Bedford Circus, to the east of the Cathedral. However, the purpose of this discussion was not to propose reinstatement of similar buildings or even the retention of the street plan. On the contrary, it is used to mobilize support for Sharp's proposals for clearly contemporary interventions, built to a new street plan. First, Bedford Circus was cited as a successful contemporary intervention of its day, an area of dramatically new architecture constructed on a new street plan. Second, Sharp argued that the popular perception of Exeter as a medieval city was misplaced and that, the Cathedral apart, the principal architectural merits of the city were Georgian. He outlined four possible forms rebuilding might take. The first was restoration, which he argued would produce a dead museum. The second and third options were a functionally modern city with medieval imagery or with eighteenth- and nineteenth-century dress; that is, a functional modern city cloaked in historic styles of architecture. He thought that these

FIGURE 8.1 Photograph of Owengate used in *Cathedral City* (Sharp, 1944).

might be popular with the public but would be contemptible. Finally, his strongly favoured option was modern renewal, sympathetic to, but not imitative of, existing forms. New development should be of a similar scale to the buildings that had been lost, and intimate rather than monumental in form. This is a clear assertion of Sharp the modernist – historic

FIGURE 8.2 'View of the proposed central improvements looking west'. Perspective by A. C. Webb in *Cathedral City*, also showing a rich town- and roofscape (original in colour) (Sharp, 1944).

reproduction is not something to be contemplated. But it is also a particular sort of modernism, organized not on historic street lines but on broader historic principles of development with the street as the principal unit of urban development and street buildings aiming to be quiet and polite and subservient to the major organizing urban landmarks, such as the Cathedral.

A new shopping street Sharp proposed in the area of Bedford Circus, Princesshay, was constructed (although the area has been redeveloped again in recent years). It was formulated to capture a historically new view of the Cathedral. In Exeter the Cathedral sits tight within an enclosed Cathedral close right in the centre of the city. While Sharp celebrated this, and opposed any ideas of a general 'opening out', Princesshay was one of five new views of the Cathedral he sought to create (see figure 8.3). Thus, in addition to being a piece of modern street-based urbanism, the redevelopment of the Bedford Circus area would contribute to an evolving picturesque townscape.

In his unpublished autobiography *Chronicles of Failure,* Sharp saw Exeter as a significant stepping stone for ideas that had been developing in his Durham plan:

> Planners were now being encouraged by the government to plan in a bold 'heroic' way – 'make no small plans', it was said. This suited my temperament well enough: but I demurred from making the kind of plans which had hitherto been regarded as normal, the more or less standardised plans in which the same kind of principles were applied, whatever the nature of the town or city they were applied to. For Exeter . . . I conceived the proper planning to be through a proper recognition of the 'genius loci', (the special and distinctive spirit of place), as I had done at Durham, and to develop the concept of 'townscape', especially of 'kinetic townscape', the progress and unfolding of urban scenes which I had apprehended there.[31]

FIGURE 8.3 Princesshay, Exeter, built broadly following Sharp's proposal in *Exeter Phoenix*. Now demolished. Photograph by the author.

It is notable that it was in the Exeter plan that he had first introduced the device of an introductory section (where he had also used the idea of *genius loci* and sought to establish the character of Exeter) and a tailpiece (mostly concerned with townscape principles, see figure 8.4). These devices were subsequently reprised in his Oxford plan, outwith the main planning proposals.

Though *Oxford Replanned* was described as 'largely a work of preservation',[32] it was, like Durham, largely based on an appreciation of the city's visual qualities. The emphasis was firmly on the appearance of Oxford and the character that derived from the way space was used, rather than the fabric of historic buildings. The visual relationships of Oxford and the progression through space were studied in great depth, and it was in this report that Sharp first used the term 'townscape'. This analysis extended beyond the major set-pieces to include, for example, 'the Backs'.[33] Many of the college buildings in particular already had statutory recognition as scheduled ancient monuments. Sharp identified further buildings of architectural value. His attitude towards relatively modest buildings was variable: Ship Street and Beaumont Street were regarded as important, whereas he was dismissive of St John's Street. The differentiated approach was based on Sharp's analysis of the contribution of each of these streets to Oxford as a whole. St John's Street was part of an area generally proposed for redevelopment and was expendable. Beaumont Street was regarded as important for retention as being unusually formal for Oxford. Ship Street, as well as being picturesque, was a foil between college buildings and Cornmarket Street. The issue of foils was stressed in the townscape principles presented in the introduction to the plan, called its 'Frontispiece'. The special importance of Oxford lay in 'the existence of the great concentration of large-unit buildings there (i.e. the colleges) . . . [but] without the Foils to the big blocks of the colleges which it possesses in such

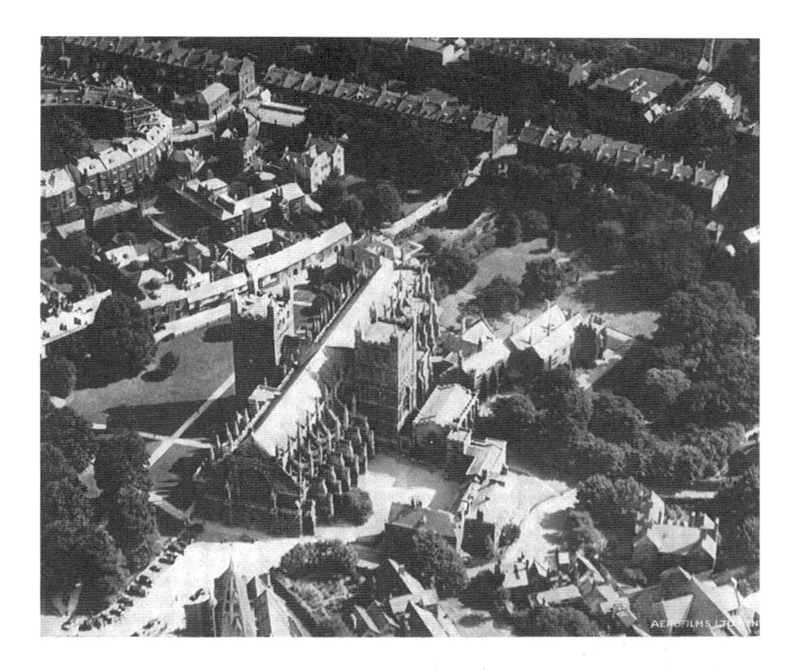

FIGURE 8.4 Page 140 from *Exeter Phoenix*, from the 'Tailpiece', discussing the importance of enclosure (Sharp 1946).

streets as Bath Place . . . Turl Street . . . and Ship Street, Oxford would be infinitely poorer. To pull down Ship Street, for example, and put up a great single-unit building there, as has been authoritatively proposed, would be a bigger architectural crime than to pull down one of the old colleges themselves.'[34] (See figure 8.5.)

Thus the emphasis east of Carfax was firmly on preservation, not only of buildings but also of a pleasant collegiate atmosphere. A key priority was the removal of traffic from the High

FIGURE 8.5 Photograph of Turl Street from *Oxford Replanned*; one of Sharp's 'foil' streets (Sharp, 1948).

Street, the principal route to the east that bisects the main college area. Sharp considered the High Street a 'great and homogenous work of art',[35] not because of the intrinsic quality of the buildings but because of the relationships between them, although he argued it was rarely able to be enjoyed, as a result of the volume of motor traffic. Thus, to facilitate this enjoyment, drastic interventions were seen as necessary. Sharp's plan was to run a road around the south of the centre through the Meadows along the line of Broad Walk, a route he termed 'Merton Mall'. Deliberations over a series of competing schemes to relieve inner Oxford of traffic were made in a series of highly controversial public inquiries between the 1950s and 1970s, a period through which Sharp remained personally involved, partly because he had moved to live in Oxford. Ultimately, most of the suggested road schemes were abandoned in favour of traffic management.[36]

In my view *Oxford Replanned* lays claim to be Sharp's defining work, at least in terms of his practice as a planning consultant. It was a major and pioneering piece of townscape analysis, reprised in *Oxford Observed*.[37] The plan's frontispiece and tailpiece effectively set out principles and components of townscape, using Oxford as an example. Together they totalled some 65 pages of analysis. As well as the High Street, Sharp analysed the nearby sequence of Bodleian Library, Radcliffe Camera and St Mary's church, finding it 'a first class aesthetic experience . . . to be treated with awe'.[38] Crucial to this experience was the experience of movement through space, one of his fundamental townscape principles (see figure 8.6). His exposition of townscape also encompassed much humbler elements. For example, under the heading of 'trivia' he considered the importance of floorscape, demonstrating the significance of texture, and its erosion through tarmac resurfacing.

It was also at Oxford that Sharp came most directly into contact with the developing manifesto of the *Architectural Review* and its co-owner/co-editor H. de Cronin Hastings. Hastings had developed a close personal interest in the plans Sharp was publishing through the Architectural Press (which Hastings also co-owned). Hastings and Sharp spent time together at Oxford, and indeed in *Chronicles of Failure,* Sharp stated that some of the text in the frontispiece and tailpiece was introduced anonymously by Hastings. Recent scholarship has argued that the townscape programme developed by the *Architectural Review* was a polemic for a modern English urbanism,[39] and at this point we can see this harmonizing with Sharp's own goals.

Whilst it is possible to observe the principles that Sharp espoused in his other published planning reports of the period, they are never as clearly articulated as in the three plans that have been discussed. The best produced of the rest was the final report for Salisbury, published by Architectural Press.[40] As a planned town essentially on a grid pattern, Salisbury did not present the complex picturesque effects which so excited Sharp in Durham and Oxford. However, he was not unappreciative of the merits of the wider town, terming it the 'most "medieval" of all English cities',[41] due to the survival of many small buildings of earlier periods (see figure 8.7). Sharp again argued for good new contemporary architecture rather than reproduction of historic styles and once more presented extensive road proposals. Of the other historic town reports, Chichester[42] was perhaps the most significant. Sharp had high regard for Chichester, describing it alongside Lichfield and Wells as the epitome of the English 'cathedral-city'. Its even greater quality, however, was his favoured urban form of high-quality Georgian architecture incrementally evolved on an older (Roman) street plan – 'the least spoiled example now remaining in England of a naturally-grown as distinct from a deliberately planned renaissance town'.[43] Sharp savaged proposals for incremental street widening through setbacks in redevelopment, promoted by

FIGURE 8.6 Kinetic townscape; the sequence of Bodleian Library, Radcliffe Camera and St Mary's church from *Oxford Replanned* (Sharp, 1948).

the County Council for the main historic road axes in the city – perhaps his most enduring contribution to the planning of Chichester.[44] Sharp characterized Taunton[45] as mostly pleasant, though not spectacular: at heart a country town both visually and functionally. He remarked that while there were a good number of buildings of architectural or historic

FIGURE 8.7 Townscape analysis from *Newer Sarum: A Plan for Salisbury*. 'In addition to a variety of colour and form, a variety in the height of buildings, with a consequent diversity of skyline, plays an important part in producing Salisbury's essentially informal character' (p. 70). The picture is of St John's Street (Sharp, 1949).

interest, the physical attractiveness of the town derived more from a good stock of humbler buildings – essentially decent, fairly ordinary Georgian. Road proposals were a significant feature of his plan. Particularly at issue in Taunton were the east-west routes towards Devon, which demanded external bypasses. However, he also suggested inner 'substitute roads' to relieve central congestion. None of Sharp's other reports were published, although some, such as King's Lynn, were extensively summarized in the professional press.[46]

It is perhaps disappointing that we do not have the counterpoint of a Durham, Exeter or Oxford for a non-historic city. The plan for Todmorden[47] was short and functional, with a principal focus on housing conditions and economic decline. From a design point of view the most interesting feature of the town was its topography, lying at the junction of three narrow valleys which quickly rise to high moorlands. Sharp's report on Stockport[48] was based around a functional reconstruction plan for the town centre. Major interventions were proposed but not a clean sweep – it was seen as desirable to maintain something of the historical pattern of the town. The crossing of north-south and east-west arterial roads was seen as a major planning issue, as was the dramatic topography. His report for Minehead[49] was not a full reconstruction plan but a report focused on the amenities of the coast and sea front. Sharp's objective was to improve Minehead as a comparatively quiet but well-equipped holiday resort. It stressed the picturesque qualities of old Minehead, much of which could be used for holiday accommodation.

Discussion and evaluation

In the reconstruction plans of Thomas Sharp we can see the development of ideas about the organization of urban space that lead on from earlier formulations in his books and work for the Ministry of Town and Country Planning. Thus core principles, such as the importance of the street as the primary urban building block, are applied to real places and real planning problems. Furthermore, they are enriched by the study of particular places, and ideas about the urban scene get crystallized into the concept of townscape. It is possible to trace the development of townscape and allied ideas through the Durham, Exeter and Oxford plans. A clear sense of Sharp's kinetic townscape, or serial vision as Cullen would subsequently label it, can be seen in the Durham plan, although it is not specifically labelled as such, and Sharp discussed the idea that places have a character that extends beyond particular buildings. In the Exeter plan this was labelled as *genius loci,* and we see townscape principles being used in the formulation of proposals for new parts of the city. In the Oxford plan we have a masterful drawing together of the principles of townscape Sharp had been developing. It is notable that the most significant plans by Sharp exceeded their brief in various ways to become manifestos, arguing for 'a way of seeing' places. This is most obvious with Exeter and Oxford, with their front and back sections disconnected from the practical planning proposals of the main plan.

As a specific adjunct to this, in the Durham and Oxford plans in particular we can see Sharp's distinctive contribution in providing for the massive growth of motor traffic in historic cities. Sharp's approach was quite distinct and different from the prevailing model evolving in the wake of the well-known recommendations of Tripp,[50] which generally led to a tight inner-ring road with a series of precincts internal to this. Sharp's critique was twofold. First, he criticized the solutions a mechanistic precinct approach produced, on the grounds that they were not place-specific and lacked proper analysis of the individual place in question. Second, Sharp's plans generally did not have inner-ring roads as such, because he argued that much urban traffic was internally generated and therefore could not be removed by bypasses and ring roads. Sharp used the term 'substitute road' for his approach. A substitute road was a road inserted close to the main congested streets (which might be commercial or shopping) and designed to relieve the principal street of all its traffic except that directly needing to be there; a distinctive approach later acknowledged by Colin Buchanan.[51]

Sharp's reconstruction plans (and others of the period) were responding to the forces of modernity and pressures that had been building up in the inter-war period, in particular the impact of the motor car and the political imperative of addressing housing conditions and separating incompatible neighbouring land-uses. To achieve these goals major changes to urban form were considered to be inevitable. The ideology that underpinned reconstruction plans was that this change should be addressed through comprehensive planning, rather than unregulated development and the muddled and unsatisfactory attempts at planning that had occurred up until that time.[52] It was the fate of Sharp that his brief period of success as a consultant in professional practice was largely associated with work on historic towns. And yet to consider Sharp's contribution in this light is to marginalize him. Whilst it may have been Sharp's special skill to apply his ideas to historic towns, he was advocating a form of urbanism that he considered relevant to *all* towns and cities. Although historic cities were excellent demonstrations of how townscape principles created great places, these ideas should, he believed, be applied in the formation of new places, whether they be modifications of existing towns or entire new settlements. Furthermore, townscape principles in turn were intended as only one part, the visual expression, of a wider urbanism – a practical English and modern urbanism, that could be implemented and used as part of a process of realizable planning, by contrast with the rather more theoretical imports arriving from the Continent. As such it formed part of a wider comprehensive planning and was not just concerned with the architectonic form of place – the objective was to synthesize functional requirements, the proper workings of the town, with visual seemliness.[53] Again, this was a mission shared with the *Architectural Review* and might be argued to be in the tradition of an atheoretical English empiricism.

Decoding the extent and duration of the influence exercised by Sharp's reconstruction plans is complex and beyond the scope of this chapter. However, he certainly had an impact on the places for which he wrote plans. While his plans were never implemented as such, one can trace his ideas in some of the ways the cities subsequently developed, either through the implementation of specific proposals or in a more general influence on the discourse of place. At a broader level, he highlighted the importance of understanding the distinctive qualities of particular places, and mobilizing the concept of *genius loci*. This was a significant influence on the subsequent conservation movement, where ideas of townscape, informed by Sharp and the propaganda of the *Architectural Review,* were to prove enormously influential in developing more holistic conceptions of place.

Notes

1. Research on this extraordinarily dynamic period in British planning has generated a large academic literature in recent years. What follows are examples only: Hasegawa, J. (1999) 'The rise and fall of radical reconstruction in 1940s Britain', *Twentieth Century British History,* 10(2): 137–161; Larkham, P. J. (2003) 'The place of urban conservation in the UK reconstruction plans of 1942–1952', *Planning Perspectives,* 18(3): 295–324; Larkham, P. J. and Lilley, K. D. (2003) 'Plans, planners and city images: place promotion and civic boosterism in British reconstruction planning', *Urban History,* 30(2): 183–205; Pendlebury, J. (2003) 'Planning the historic city: 1940s reconstruction plans in Britain', *Town Planning Review,* 74(4): 371–393; Tiratsoo, N. (2000) 'The reconstruction of blitzed British cities, 1945–55: myths and reality', *Contemporary British History,* 14(1): 27–44.
2. Pendlebury, J. (2009) 'The urbanism of Thomas Sharp', *Planning Perspectives,* 24(1): 3–27.
3. Sharp, T. (1932) *Town and Countryside: Some Aspects of Urban and Rural Development,* Oxford: Oxford University Press.

4. Sharp, T. (1936) *English Panorama,* London: J. M. Dent & Sons; Sharp, T. (1940) *Town Planning,* Harmondsworth, Middlesex: Penguin; Sharp, T. (1950) *English Panorama,* 2nd edn, London: Architectural Press.
5. Williams-Ellis, C. (1928) *England and the Octopus,* Portmeirion: Golden Dragon Books.
6. Now the Campaign to Protect Rural England.
7. Sharp, *Town and Countryside,* p. 143.
8. Sharp cited D. H. Lawrence as an example of a writer who argued that English towns and cities never achieved true urbanity. See e.g. Sharp, *Town Planning.*
9. Hussey, C. (1927) *The Picturesque: Studies in a Point of View,* London: G. P. Putnam's Sons. See essay by Aitchison in Pevsner, N. (edited by Mathew Aitchison) (2010) *Visual Planning and the Picturesque,* Los Angeles: Getty Publishing, for a discussion of these issues. See also Atkinson, this volume.
10. Sharp, *Town and Countryside,* p. 140.
11. See Atkinson, this volume.
12. Sharp, T. (1968) *Town and Townscape,* London: John Murray. It is questionable how much this represented a change of view. He did not dismiss the principle of building tall; at the heart of his critique was, for him, the bad urban manners many such buildings displayed.
13. Trystan Edwards, A. (1924) *Good and Bad Manners in Architecture: An Essay on the Social Aspects of Civic Design,* London: Philip Allan. Sharp's text quite closely follows Trystan Edwards' early discussion of terrace design.
14. Cullen, G. (1961) *Townscape,* London: Architectural Press.
15. Sharp T. (1942) *Civic Design: With Special Reference to the Redevelopment of Central Urban Areas,* Unpublished manuscript in the National Archives, Kew. National Archives HLG 71/779, London, p. 29. This manual was one of a series planned for publication as guidance for local authorities, but their development and publication seem to have fallen through, due to bureaucratic indecision and delay. Sharp also produced a manual on village design which he was later, after much wrangling, allowed to publish independently as *Anatomy of the Village* (1946, Harmondsworth, Middlesex: Penguin).
16. Sharp, *Civic Design,* p. 29.
17. Ibid., p. 37.
18. Ibid.
19. See Pendlebury, 'The urbanism of Thomas Sharp', for a discussion of these issues.
20. The term 'reconstruction plan' is used here as short-hand for the advisory plans produced by or for localities in the mid- to late-1940s especially. In many of these settlements 'reconstruction' was not a response to war damage, and it is intriguing that many settlements without any major urban problems chose to commission or undertake them. See Larkham, P. J. and Pendlebury, J. (2008) 'Reconstruction planning and the small town in early post-war Britain', *Planning Perspectives,* 23(3): 291–321, for a discussion specifically on small towns.
21. Sharp, T. (*c.* 1973) *Chronicles of Failure,* GB 186 THS, Newcastle upon Tyne, p. 296.
22. Ibid.
23. Sharp, T. (1945) *Cathedral City: A Plan for Durham,* London: Architectural Press.
24. Sharp, T. (1937) 'The North East – Hills and Hells' in C. Williams-Ellis, (ed.) *Britain and the Beast.* London: Readers' Union, pp. 141–159. For a longer description of how the plan was commissioned see Larkham and Pendlebury, 'Reconstruction planning and the small town'.
25. Sharp, *Cathedral City,* p. 15.
26. Ibid., pp. 88–89.
27. Ibid., p. 53.
28. Ibid., p. 54.
29. Ibid., p. 41. Sharp's emphasis.
30. Sharp, T. (1946) *Exeter Phoenix: A Plan for Rebuilding,* London: Architectural Press.
31. Sharp, *Chronicles of Failure,* p. 236.
32. Sharp, T. (1948) *Oxford Replanned,* London: Architectural Press, p. 16.
33. 'The Backs' is a term usually associated with Cambridge. However, in this plan Sharp extends its use to areas of collegiate development around quadrangles more generally, explicitly applying it to Oxford.
34. Sharp, *Oxford Replanned,* p. 44.
35. Ibid., p. 20.
36. Stansfield, K. (1981) 'Thomas Sharp 1901–1978' in Cherry G.E. (ed.) *Pioneers in British Planning,* London: Architectural Press, pp. 150–176.

37. Sharp, T. (1952) *Oxford Observed,* London: Country Life.
38. Ibid., p. 32.
39. Erten, E. (2009) 'Thomas Sharp's collaboration with H. de C. Hastings: the formulation of townscape as urban design pedagogy', *Planning Perspectives,* 24(1): 29–49.
40. Sharp, T. (1949) *Newer Sarum: A Plan for Salisbury,* London: Architectural Press.
41. Ibid., p. 10.
42. Sharp, T. (1949) *Georgian City: A Plan for the Preservation and Improvement of Chichester,* Brighton: Southern Publishing Corporation.
43. Ibid., pp. 16–17.
44. Larkham, P. J. (2009) 'Thomas Sharp and the post-war replanning of Chichester: conflict, confusion and delay', *Planning Perspectives,* 24(1): 51–75.
45. Sharp, T. (1948) *A Plan for Taunton,* Taunton: Taunton Corporation, p. 64.
46. Sharp, T. (1948) 'King's Lynn: a redevelopment plan and some notes on the planning of the borough', *Architects' Journal* (30 December): 597–602.
47. Sharp, T. (1946) *A Plan for Todmorden,* Todmorden: Borough Council, p. 35.
48. Sharp, T. (1950) *Stockport Town Centre Replanned,* Stockport: Stockport Corporation, p. 18.
49. Sharp, T. (1950) *Minehead: The Development of Its Amenities,* Minehead: Minehead Urban District Council, p. 20.
50. Tripp, H. A. (1942) *Town Planning and Road Traffic,* London: E. Arnold.
51. Buchanan, C. (1958) *Mixed Blessing: The Motor in Britain,* London: Leonard Hill.
52. Pendlebury, J. (2004), 'Reconciling history with modernity: 1940s plans for Durham and Warwick', *Environment and Planning B: Planning and Design,* 31(3): 331–348.
53. Pendlebury, 'The urbanism of Thomas Sharp'.

9

THE ROLE OF A HISTORIC TOWNSCAPE IN CITY RECONSTRUCTION

Plans for Milan, Turin and Genoa after World War II

Francesca Bonfante and Cristina Pallini

Introduction

In 1993 the Italian architecture journal *Rassegna* devoted an issue to city reconstruction in Europe after World War II.[1] Nearly 50 years after the reconstruction period, Gregotti wrote that a new trend in reconstruction was becoming evident across Europe and due consideration should be given to the means that town planning and architecture could offer for radical intervention in urban growth.[2] Olmo drew attention to some key themes and to their exponents who had led the debate in Italy immediately after the war, indicating crucial dates and calling for further study to 'split up history into its fragments', to restore to that historic period its many and varied opinions and hopes, without which its portrayal risked simplification to the point of becoming a sterile myth.[3]

An extensive literature on the subject stressed the originality of the post-war debate in Italy, attempting to bridge the gulf between architecture and town planning.[4] According to Menghini, the Italian experience clearly showed a desire to restore the urban question to the realm of architecture, often adopting an empirical attitude far from mainstream modernism's 'theories and models' alternative to the historic city.[5] Piccinato explored the changing social and economic contexts that formed a background for debate in Italy from 1945 onwards, and emphasized the importance of the INA-Casa (Instituto Nazionale delle Assicurazioni, the National Insurance Association) plan, a programme promoted by the government to increase employment by supporting the construction of housing for labourers.[6]

In addition, Italian architects and engineers[7] – who were building not only for the indigenous population but also for migrants bringing their own cultures, although desirous of integration into the new context[8] – became increasingly conscious of their social role while drifting away from the planning principles of the Modern Movement. Realizing the importance of operating at the concrete level of a city's historic dimensions, Italian architects showed renewed interest in local building types and traditions.[9]

Referring to the period from 1945 to 1960, this chapter discusses the case studies of Milan, Turin and Genoa that together formed the so-called industrial triangle, where internal migration had been concentrated even before the two world wars. Here more than elsewhere, reconstruction had to cope with addressing the city's operational efficiency (industry and infrastructure) while also envisaging a 'cultural structure' for a new social edifice. A specific theme has been chosen for each city: planning (Milan), new residential districts (Turin), museums (Genoa) – thus covering the three dominating themes of post-war reconstruction in Italy.

While the plans were beautiful, most remained unimplemented, arousing public debate over the future of large Italian cities and often revealing a variety of approaches: some mainly focused on infrastructure to improve links between roads, railways, canals and ports; others delving into the growth of a city throughout its history, the better to identify which distinguishing features of its structure and architecture were to be preserved. Large districts of low-cost housing and new civic museums were thought vital for integrating the new urban populations: while housing districts formed part of the changing suburban landscape, museums showed how the old could relate to the new in the historic heart of the city.

Diverse in their solutions, the works selected here are representative of research carried out by a number of leading personalities who sought to combine new expressive forms with an 'ethical role' in town planning and architecture. The Milan case is focused on the debate over the new plan by engineers and architects who included Cesare Chiodi,[10] Giuseppe De Finetti,[11] Luigi Lorenzo Secchi[12] and Ernesto Nathan Rogers,[13] as well as by technical staff on the City Council. With regard to Turin, Giovanni Astengo[14] is a leading personality in understanding how the Falchera housing district can be related to the Piedmont Regional Plan. Genoa is important for its museums, newly arranged after World War II by Franco Albini:[15] these were among the best in Italy, housing civic collections which were exhibited with a view to gaining the widest possible recognition of their 'educational and cultural potential'.

Italian architects in the aftermath of war

By 1945, with widespread devastation across the country, a desire for individual and collective catharsis had led to a demand for institutional and political regeneration. At the end of the war, the cultural context had changed radically, profoundly marked by the Resistance movement. The majority of Italian architects had accepted Fascism. Before the war only a few had openly shown their feelings against the regime, but even those who had taken an active part in the Resistance became truly anti-Fascist only much later on. Writers, artists and film directors, as well as many architects and town planners, had been strongly sympathetic to the Resistance movement and were seeking an ethical renewal, at times animated by the will to understand the real political and social nature of Fascism. Architects, in particular, sought a better understanding of the nature and social composition of the people for whom houses and services were to be provided in Italy.

A cultural revival was marked by the appearance of several publications[16] and by the almost simultaneous creation of groups representing different trends: in Milan the MSA (Movimento Studi per l'Architettura),[17] in Rome the APAO (Associazione per l'Architettura Organica)[18] and in Turin the Gruppo Pagano.[19]

The period when discussion among intellectual elites was paralleled by institutional and economic reorganization can be said to have begun in Milan in December 1945, at the First National Meeting for Reconstruction,[20] and to have ended in Paris, with Italian participation at the Exposition Internationale de l'Urbanisme et Habitation,[21] held at the Grand Palais in the summer of 1947. At the Milan meeting a timely confrontation took place between the protagonists of progressive culture, the supporters of planning and the exponents of the capitalist establishment. At the Paris Exhibition 'Italy the Rebuilder' made its formal appearance on the European scene. In the five sections (*Problème du Logement, Urbanisme, Habitation, Construction, Information*), alongside plans for rebuilding the great cities and some important historical centres, Italy presented a draft of the Plan for Milan, the Piedmont Regional Plan and experimental designs for new neighbourhood units. The measures for reconstruction which Italy – with its age-old traditions of building – was taking, or proposing to take, were considered of particular interest to all other countries, whether closely or distantly affected by similar problems. The catalogue of the exhibition presented Italy as a symbolic case because of the gravity of its housing problem and the critical condition of its people, its infrastructure and all of its productive resources. A few months later, the elections of 18 April 1948 excluded left-wing representatives from participation in the government and from strategic decisions, marking the demise of any 'revolutionary' prospect. This was followed by a period of economic neoliberalism destined to continue for many years.

City reconstruction was subjected to plans formalized by a decree of March 1945.[22] While the Town Planning Act of August 1942[23] required that each Master Plan be linked to a regional prospect, reconstruction plans were concerned only with destroyed or severely damaged urban areas. Working with no overall idea of the national reconstruction (unrelated to the general picture of local urban problems), failing to face the question of public ownership of land and showing little respect for characteristic environmental forms, reconstruction plans often produced structures inferior to those they replaced. These plans often only made previously existing situations worse, due to the urgency of their implementation and the possibilities for putting them into practice straight away.

Visions of a future Milan

Milan was certainly a unique case in Italy, not only for its economic and cultural importance, but also for the many and timely proposals in its plan.[24] According to Alberto Mioni, Milan was an exceptional case for its lively and exhaustive debate on the future of the city, involving professionals as well as ordinary people, despite the fact that (in his opinion) nearly all the proposals included ideas that had characterized the Fascist period, when Milan was seen as an 'ideal middle-class city', occupying a dominant position in the region of Lombardy.[25] In our view these early proposals showed alternative visions of post-war reconstruction.

Mioni discussed these early proposals, pointing out the differences between the 'plan' and 'counter-plans'. He considered the 'plan' as being a revision of the 1934 plan by Secchi, presented at the town planning section of the Engineering and Design department of the Milan City Council (but cancelled by the National Liberation Committee[26] because of its associations with the Fascist administration). The three counter-plans were, respectively, those of a group of Milanese engineers under the guidance of Chiodi, the plan by AR (Reunited Architects) and the proposals by De Finetti. It may be noted that the two groups of engineers and architects both attributed overriding importance to deciding which kind of infrastructure would best favour decentralization of the population and industrial activity. The 'plan' by Secchi and the 'counter-plan' by De Finetti, on the other hand, were three-dimensional and included ideas on how certain buildings (old or new) and public spaces could form a composition at the urban level that linked the future city with existing features of its long history, inventing spatial frameworks of a new kind.

The proposal by AR and ideas from the Engineers' Union

Begun in early 1944 by a group of architects inspired by the principles of the Charte d'Athènes, the AR plan[27] (Fig. 9.1) envisaged Milan as a vital part of 'a new Italian and European style of living'[28] whose hierarchically organized accessibility allowed for an urban population of 800,000, with 200,000 living in the suburbs and surrounding area (up to 40km from the city centre). To avoid the monocentric ring shape of previous plans, this adopted a historic NW-SE / NE-SW orientation of the two main highways, crossing at right angles near the old city centre. The highway was to connect the city's functional core: the goods station, the old centre (to become a residential area free of heavy traffic and surrounded by parks and gardens), the Trade Fair and adjacent business district,[29] the sports area and Malpensa airport; the proposed canal and the railway sidings were to attract industrial complexes. Conceived as the key to regional development, the AR plan envisaged, first and foremost, the decentralization of industry and the creation of a new central business district north of the old centre – ideas that were to be embodied in the final version of the Master Plan, which was approved on 30 May 1953.[30]

FIGURE 9.1 AR Plan for Milan, 1944 (F. Albini, BBPR, P. Bottoni, E. Cerutti, I. Gardella, G. Mucchi, M. Pucci, A. Putelli).

Studies for rebuilding Milan had also been promoted since March 1944 by the Engineers' Union; coordinated by Chiodi, these studies benefited greatly from a lively public debate.[31] Chiodi believed that town planning should do away with questions of form and deal with problems of resource management and public policies, addressing aspects of demography and land values, building and finance. The main idea put forward by this plan concerned the decentralization of the urban population and industrial activity, to be achieved by establishing an efficient network of public transport including new underground lines and the canal. This idea of decentralization was also to be pursued by fostering forms of 'integrated economy' (industry and agriculture) peculiar to the areas around Milan.

The 'plan' by Luigi Lorenzo Secchi

While Mioni considered the Secchi plan to be a point of reference for what he termed the 'counter-plans', Morandi called it a 'phantom plan', sidelined yet ever present.[32] It should be remembered that Secchi had worked in the Engineering and Design department of Milan City Council since 1926, playing a part in decisions concerning where and how to build a large number of sports facilities and schools. The origin of his plan for Milan can be traced to the studies he began in 1938, which aimed to correct the 1934 plan.[33] In the winter of 1943, after the heavy bombing of civil targets, Secchi was asked to abandon partial solutions and instead draw up a new plan for the whole municipal area. A first draft of the Secchi plan, with diagrams and drawings, was discussed in April 1944 and completed a year later in March 1945. Two large tempera-painted drawings were put on public display at the Milan Triennale: one of the central area included within the Spanish ramparts (demolition of which, begun in 1884, was completed after World War II), the other of the outer area.

Believing that the city's future prospects were to some extent independent of the national political orientation, Secchi estimated that the population of Milan would increase to 2,250,000 in 50 years. His zoning scheme included green areas, noticeably reducing the areas amenable for urbanization, all in contrast with the ring form of previous plans. Secchi envisaged a tentacular urban form, proposing that development be concentrated to the north outwards along the main arterial roads, separated by vast 'green wedges'; while in a southward direction, development was to be limited to allow for the presence of springs, a close network of irrigation ditches and extensive water meadows. For infrastructure, his plan attempted to establish some degree of consistency among projects already begun or discussed: the canal to Cremona along the Po river, a redesigned railway junction and new sidings, the underground, a system of ring roads.

Secchi's original contribution lay in his proposals for the city centre, devastated by the bombing. The number of new roads across the centre was drastically reduced, as was much of the demolition planned prior to the war. Coexistence of old with new was seen not as a problem of style but rather one of volumetric and spatial relations. Both led to the idea of a three-dimensional plan; preserving the face of the city and saving what could be saved of the historic and architectural features in the centre 'the existing town plan was studied not merely to offer a planimetric solution for the environment, but rather to find a harmonious balance between plan, heights and volumes'.[34] To remedy the artistic and monumental impoverishment of the city centre, Secchi proposed a network of 'heritage walks' linking small wooded areas, private and historic gardens open to the public, areas that had become vacant due to the bombing or to the decentralization of public buildings (Fig. 9.2). Green spaces were seen as forming a vital part of urban reconstruction. Passers-by were intended to rediscover a series

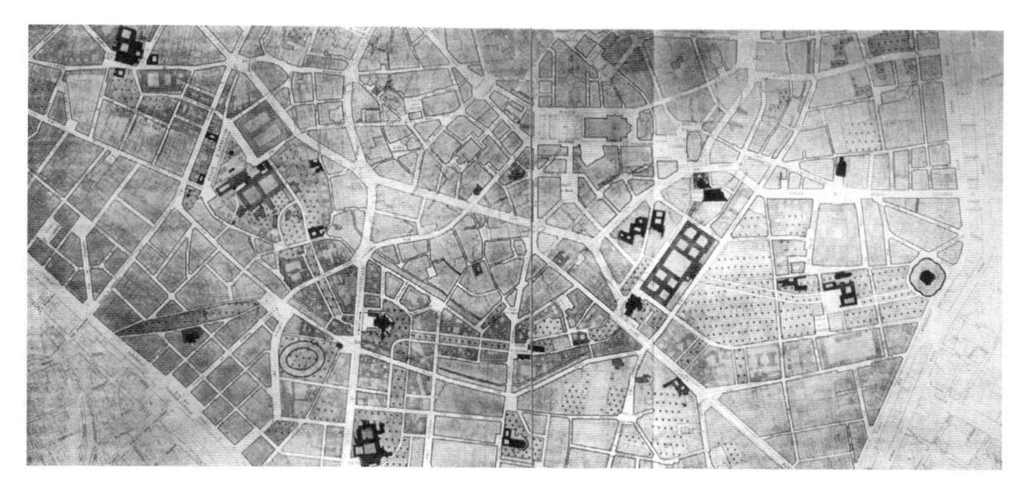

FIGURE 9.2 Heritage walks in the centre of Milan, 1945 (L. Secchi) (courtesy of Secchi-Tarugi family archive, Milan).

of historical settings, sites of the city's most ancient monuments: the Basilica of St Ambrogio (379–86 AD), the Basilica of San Lorenzo (*c.* 370 AD), the Basilica of St Eustorgio (founded in the fourth century), vestiges of the Roman Arena, the church of St Stefano (1643–74) and the Ca' Granda Hospital (begun in the fifteenth century).

The 'counter-plan' by Giuseppe De Finetti

Before the war De Finetti had proposed a number of projects in opposition to the 1934 plan. Later, from March to September 1944, he took an active part in the conferences promoted by the Engineers' Union.

In November 1945 he entered the competition for a new Master Plan. Excluded from the team elected to draw up the plan, De Finetti believed that reconstruction required innovation in planning procedures and legislation, and he criticized the Milan administration on a number of grounds: the poor level to which recollection of the city had sunk, secrecy in public affairs, a lack of data needed for proper estimation of the tasks ahead and the survival of ideas belonging to the pre-war period – 'gigantism, imperialism, nationalism, parochialism'.[35] His suggestions for a new layout were based on detailed historical study, as he explained in *Milano Risorge.*[36] Believing that the plan then under consideration overestimated the city's functional aspects, De Finetti considered Milan as a city closely integrated into its region, a physiological dualism (*forma urbis* and *natura agri*) that could never be fully expressed by any approach to planning restricted to the collection of data and the application of standards.

In opposing the main ideas of the new Master Plan – a central business district close to the main highway intersection, housing development through neighbourhood units – De Finetti identified a number of 'key themes' linked to the new accessibility and relations to a context well outside the city itself. Their nature was that of so-called typological devices that may be grouped into two main 'families': one including projects for the old city centre, the other projects for building a new one.

After the war, De Finetti observed, Milan seemed like a living body wounded to its very heart. While recognizing the role of the service sector in the city's future, he proposed to revitalize the ancient trading centre around the Cathedral Square (Piazza del Duomo), offering a valid alternative to the business district. A key element of this vision was to be the 'Strada Lombarda' (1944–46), an open-air passageway 170m long, extending the shorter arcades of the Galleria Vittorio Emanuele II on either side to provide a connecting artery for the satellite streets and squares (Fig. 9.3). On the basis of an economic analysis of land profits for the central areas, where no other form of land use could compete with business uses, De Finetti proposed low-rise commercial buildings with easy public access. He opposed high-rise buildings in historic centres as being an insult to Italian cities and a bad investment, confusing the value of areas as providers of rent income with their other values as 'attraction to the public'.[37]

Grafted onto a parallel row of commercial buildings, the Strada Lombarda was able to accommodate different types of traffic. Public transport was installed underground, with escalators connecting with the ground level, while the upper loggias gave a fine view of the Cathedral and adjacent monuments. For the area around the Cathedral Square (1944–52), De Finetti used architecture to solve problems of accessibility, proposing that different levels be available to the public through a system of underground squares with ramps to the parking facilities, also underground.

As early as 1933 De Finetti had proposed a new centre for Milan in his design for converting the neoclassical Arena Civica into a much larger stadium. After the war, from 1946 to 1951, as representative of the Provincial Delegation on the management committee of the Trade Fair, De Finetti again raised the question of a new city centre, collaborating on a plan for rebuilding the Trade Fair and surrounding area. His proposal embodied a 'vision' of the whole area NW of the historic centre (Fig. 9.4). His V-shaped plan consisted of lines diverging from the old centre, these being the main road axes parallel to the Olona (NW) and Lambro (NE) rivers, where textile, steel and engineering industries had been concentrated in the nineteenth and early twentieth century. His project revived the main NW road as a sort of modern *cardo*[38] grafted onto the old centre (particularly the urban complex formed by the Park, Sforza Castle and Foro Bonaparte). The *cardo* was to orient a new street grid, while connecting an extension of the Trade Fair. Subsequent stages of expansion

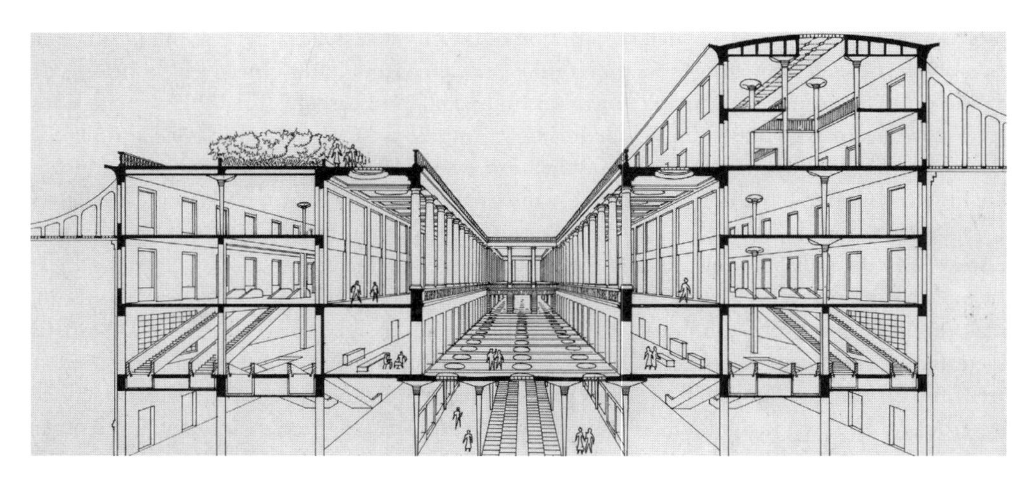

FIGURE 9.3 'Strada Lombarda', 1944–46 (G. De Finetti).

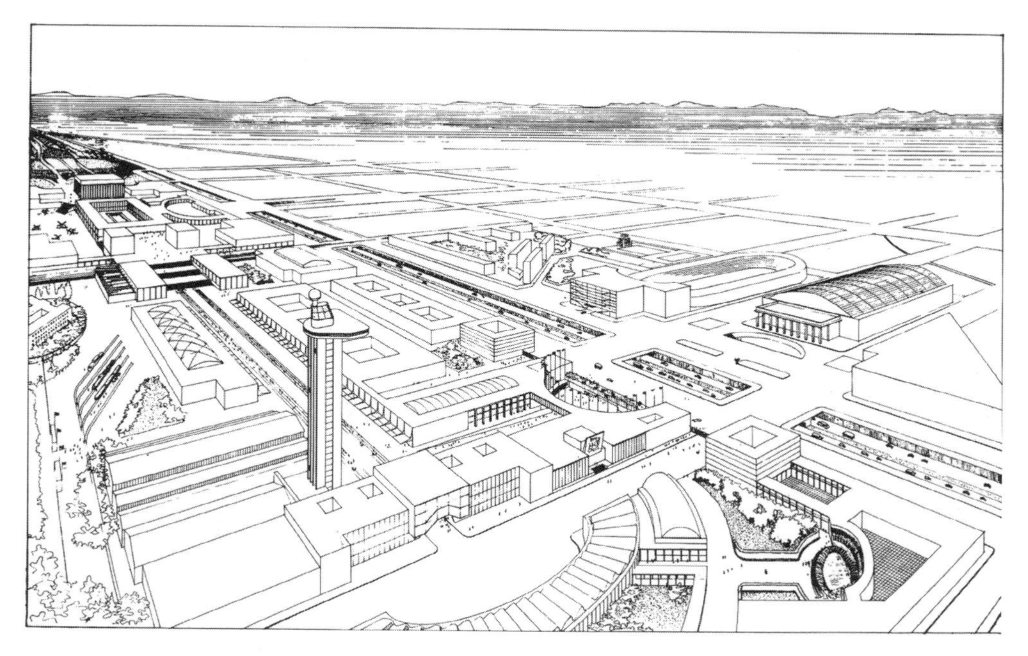

FIGURE 9.4 View of the Trade Fair area, 1943–51 (G. De Finetti).

would also require connections to the motorway and rail networks or to a heliport, but also better links with the town as a whole, with services and innovative facilities. With its great pavilions for housing temporary and permanent activities (exhibitions, shows, cultural and sports events), the Fair district with its new street grid could assume the role of 'Acropolis' to the whole city.

Turin developed to focus on a single company

By the end of the nineteenth century Turin had already become established as an industrial city, and a few decades later car production by Fiat came to dominate the urban economy;[39] even before the war, the historic grid street layout was surrounded by a ring of industrial districts attracting both regional and extra-regional migration. During the latter months of the war, after the destruction caused by the Allied bombing of 1943 (when 46% of the industrial plant was destroyed), studies were initiated in Turin on a plan for the whole Piedmont Region, an operation which exemplified the variety of factors to be tackled for the comprehensive planning of large and complex regions. Believing that this might become 'the plan of all plans in which individual programmes formed part of a single picture',[40] Astengo further developed the method and content of the Piedmont Regional Plan, collecting analytical data on the current conditions, including the nature of the region and its population, industry and agriculture, and road and rail infrastructure. Astengo sought to solve the problems of overall reconstruction to be put into effect within a defined period of time.[41] The future structure of Turin was to play a crucial part in this scenario.[42] With a view to restoring its former prosperity as a city gravitating around a single company, Astengo gave priority to transport infrastructure, of vital importance for strengthening relations between the city and its region, but also for deciding which roads could sustain decentralized industrial plant and adjacent working-class districts. Crossed by a N-S road for through traffic from and to Milan, Genoa, Savona and the Alpine passes – which also linked the city's historic core with present, and future, industrial centres at Mirafiori[43] (south) and Stura-Settimo (north) – Turin was envisaged as a great, 'organically conceived' industrial town (Fig. 9.5). Borrowing the idea of a 'neighbourhood unit' from contemporary experience in Europe,[44] the housing question was seen as an opportunity for promoting new centres of collective life. In this way the residential part of the town would grow as a federation of independent units. On leaving the centre, the N-S thoroughfare was to reach the Turin-Milan motorway and run parallel to the railway and to the proposed navigable canal. This would usher into being the initial stretch of the 'productive strip of the Po valley', an infrastructure that was to attract industry and related housing districts with a view to absorbing part of the local and national population in excess of requirements in some fields of production.

One of the perspectives published in *Metron*[45] shows part of this strip extending from Stura to Settimo Torinese. The view is taken from the top of Soperga hill (the same view extolled by famous travellers including Rousseau and De Amicis): the magnificent range of the Alps outlined in the distance, the confluence of the Stura di Lanzo with the Po river making a natural boundary at the foot of the hill, the flat countryside dotted with farmhouses and rural settlements (Fig. 9.6). The new strip appeared as a sort of linear city with an industrial zone extending from the motorway to the proposed navigable canal with its docks, ramps and loading and unloading areas. Residential units on the opposite side of the motorway were separated by large green spaces.

Astengo reinterpreted this urban context, framed in its surroundings,[46] in his project for the Falchera housing district (Fig. 9.7). Writing at length on the subject[47] he explained that this new 'organic' neighbourhood, which was to house 6,000 people, was located between Stura and Settimo where he had planned the 'new linear unit'. Referring to contemporary European experiences (borrowed from Abercrombie and Mumford among others), Astengo considered the 'neighbourhood policy' to be a first move in addressing the housing problem rather than leaving it in the hands of speculators. His approach may also be traced back to Adriano Olivetti's idea of 'community'.[48]

The Quartiere Falchera project was part of the INA-Casa plan for working-class housing (from the first seven-year period, 1949–55), requiring designers of each housing district to take into account the condition of the ground, amount of sun, landscape, vegetation, former environment and 'sense of colour',[49] and to evaluate various urban compositions and reduce causes of friction among neighbours by limiting the number of floors and occupants as well as by varying the layout.[50] These housing districts were to be built in municipalities already able to provide services, or in areas adjoining work places, adopting a village-type layout around a square close to the main social functions (church, schools, shops), with residential units around large courtyards open to the public.

The Quartiere Falchera was planned with little or no knowledge of who its future inhabitants would be (it was presumed that the size of families would be above average, and that the density would be 200 people per hectare). To create the sensation of 'another type of town, another way of living', the designers sought to strike a balance between the 'rural aspect' of the site and its urban character, a reinterpretation of the open-courtyard style of the nearby farms, with low buildings spaced out to permit a view of the Alps, the Turin 'hill' and the surrounding countryside. The long, shady roads leading to the residential blocks embraced extensive gardens offset by small squares that were to contain the administrative functions of social life.

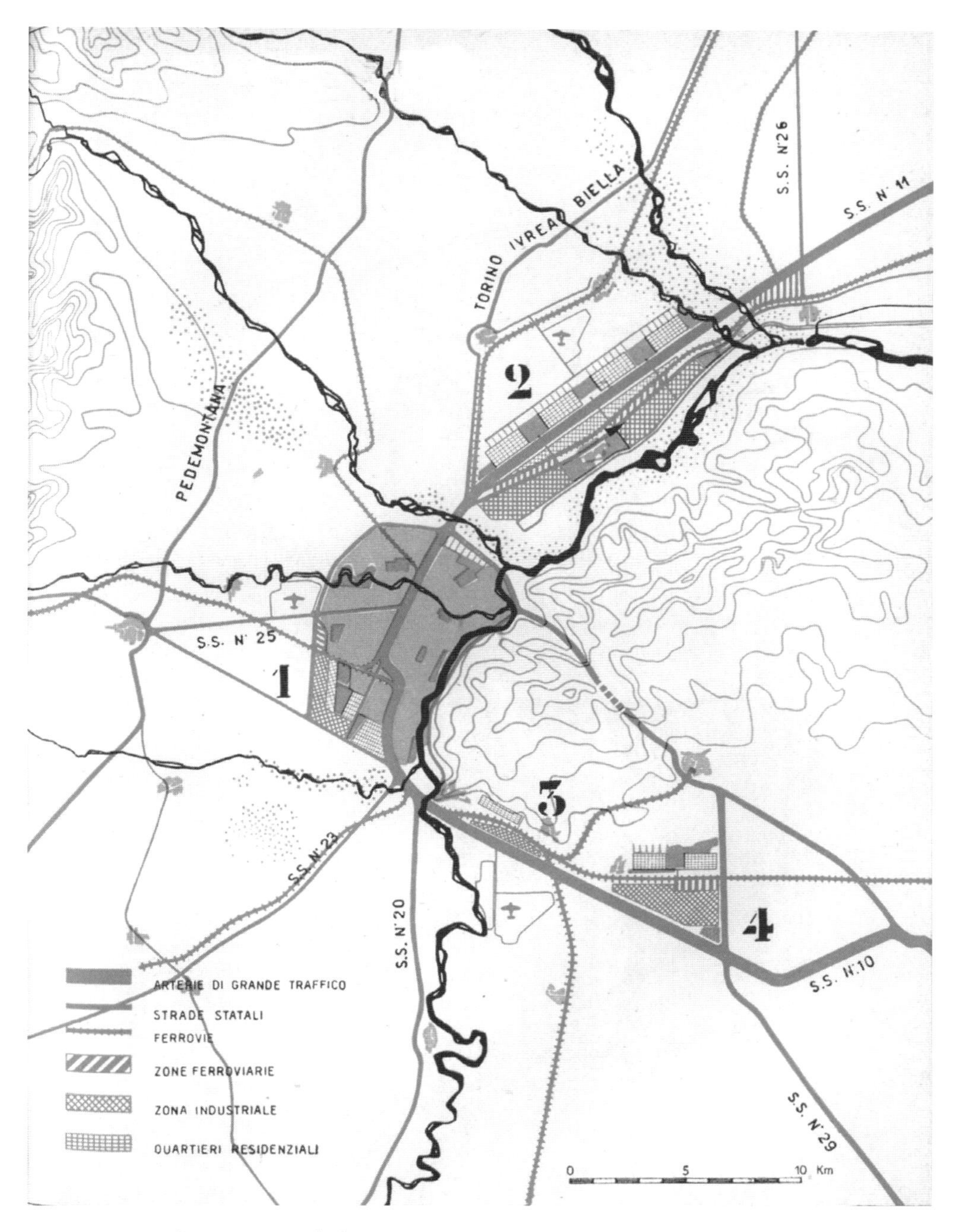

FIGURE 9.5 Piedmont Regional Plan, 1944–45 (G. Astengo, M. Bianco, N. Renacco, A. Rizzotti).

The first residents arrived in the summer of 1954. Unexpectedly, however, these were not Turin office workers from the nearby industries, but migrants from the country's southern regions or from the Friuli Venezia-Giulia region, with their own traditions and ways of life, often regretting having left their homes and reluctant to adapt to their new ones.

FIGURE 9.6 Piedmont Regional Plan: 'productive strip' from Stura to Settimo Torinese, 1944–45 (G. Astengo, M. Bianco, N. Renacco, A. Rizzotti).

FIGURE 9.7 Quartiere Falchera at Turin, 1951 (G. Astengo, S. Milli-Boffa, M. Passanti, N. Renacco, A. Rizzotti).

Genoa: old buildings for new museums

'In a country like Italy where civilization has remained creative over the centuries, the scars bearing witness to the past are added to those of the living nation.'[51] The explanatory text, published in the catalogue of the Paris Exhibition, introduced the issue of relating the new to the old that arose in most Italian cities, with their historic monuments and individual townscapes. The question of how to build a city of the future without losing the existing architectural and environmental heritage was well expressed by Rogers at the eighth CIAM (Congrès Internationaux d'Architecture Moderne), held at Hoddesdon (UK). Rogers spoke of the need to preserve, move, re-establish, enliven or reinvent the heart of a city in different places and circumstances: 'the composition of a complete work, though logical and elegant, cannot fulfil the set aims if it does not also achieve a rich, varied and surprising orchestration'.[52] As editor of the new series of *Casabella-Continuità,* Rogers initiated a review of the heritage left by the masters of the Modern Movement[53] (whose rules had since degenerated into the formalism of the International Style), trusting in a renewed relationship with history and national traditions, adapted to the environmental features in any urban area.

Within the context of this debate, the design for new museums became a 'dominating theme'. In 1949 Argan declared his hopes for a comprehensive reform of Italian museums. Recalling that the creation of public museums went hand in hand with recognition of the educational capacity of the arts, Argan borrowed Dewey's idea of continuity between art and life, believing that museums should attract and educate the widest possible public, thereby not only attracting admiration to works of art, but making them become part of community life: 'only the formal type of education that we receive from art enables us to situate the actions of daily life in a certain time and space, to learn to understand the world in which we live and work'.[54]

A similar idea of the museum as a vital place (and institution) for cultural reconstruction also emerged from a lecture by Albini at the Venice Institute of Architecture: 'Architecture must mediate between public and exhibits, must exert its powerful influence on the visitor's perception. . . . In my view it is the voids that need to be created, air and light being the building materials'.[55] In collaboration with Caterina Marcenaro,[56] Albini designed three museums in the historic heart of Genoa: the Palazzo Bianco gallery (1949–51), the Palazzo Rosso civic galleries (1952–62) and the Museo del Tesoro at the Cathedral of San Lorenzo (1952–56).

A major Mediterranean port since the Middle Ages, Genoa had been badly bombed during World War II. Its port and industries, and the densely populated historic centre known for its monuments and panoramas, were damaged. Studies for the Master Plan were begun in 1945, setting a priority on reorganizing the port and rail infrastructure and on improving accessibility by road. A concept of total respect was adopted for the historic centre, mention being made of a proposal to unite the existing parks with their fine views over the natural amphitheatre formed by the port. Several successive competitions were announced for rebuilding the damaged parts of the centre.

While aiming at closer links between city, society and culture, Genoa's museums, reordered by Albini, show an attempt to confer a new role on historic townscapes and complexes situated close to areas that for decades had been subjected to the attentions of speculators.

Both Palazzo Bianco and Palazzo Rosso are located along the Corso Garibaldi, also known as Strada Nuova dei Palazzi, an outstanding example of sixteenth-century town planning[57] that enabled Genoa to celebrate her leading role in international finance at that time. The Museo del Tesoro at the Cathedral of San Lorenzo recalls the golden age of the Most Serene Republic of Genoa as an unchallenged maritime power. Contrary to the alleged spatial neutrality of nineteenth-century museums, these three were laid out as a physical and metaphorical itinerary to rediscover both the museums and their locality, restored to a new life after wartime destruction. The need to renew the buildings that housed them, often of historical importance, was allied to the political choice of making a symbolic attribution expressive of rediscovered

FIGURE 9.8 Palazzo Bianco Gallery at Genoa, 1949–51 (F. Albini).

FIGURE 9.9 Palazzo Rosso Galleries at Genoa, 1952–62 (F. Albini).

civil values.[58] In line with Marcenaro's idea, Albini sought to overcome the concept of history as an aesthetic stratification of documents and objects, and to explore the didactic potential of the museum, its function in 'visual education' with the capacity to widen the experience of the greatest possible number of people. In this way the expository space became a theatre of experimentation in design, making possible avant-garde choices in relating the antique to the new.

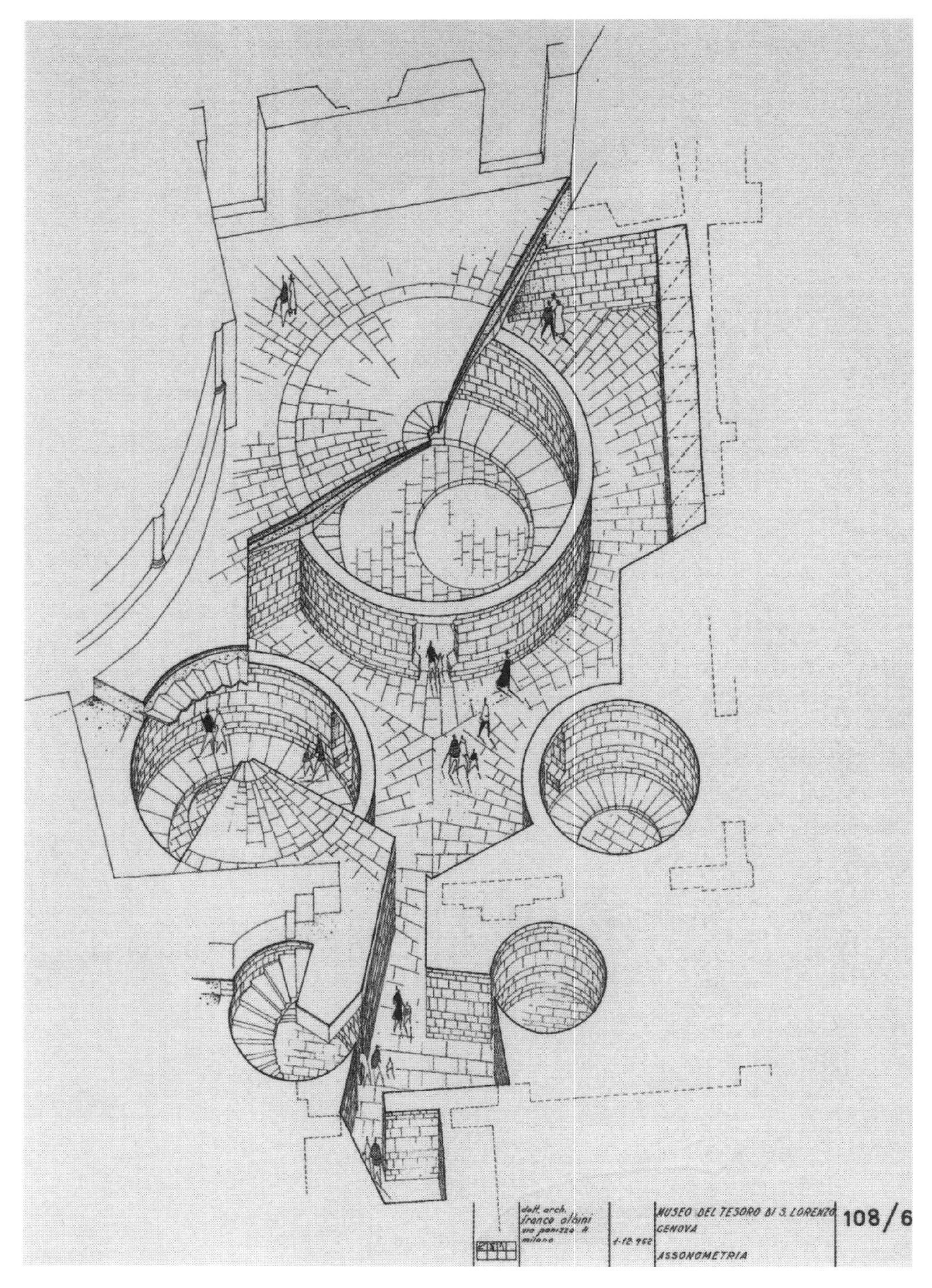

FIGURE 9.10 Museum of the Tesoro di San Lorenzo at Genoa, 1952–56 (F. Albini).

The restoration of the richly decorated Palazzo Bianco (Fig. 9.8) and Palazzo Rosso (Fig. 9.9) contrasted with the plain and simple modernity of Albini's arrangements exemplifying Marcenaro's concept of a museum: a rigorous selection of works for display in chronological order or by linguistic affinity. Whether surrounded by the severe architecture of Palazzo Bianco or by the baroque-style decorations of Palazzo Rosso, with his sparing use of materials and means, Albini achieved a high degree of formal purity. Paintings were hung from short iron bars sliding on iron guides, or else were displayed on stylized easels standing on antique stone bases; flagstaff-style supports were fitted with telescopic horizontal arms of different sizes according to the size of the painting; items of applied art were shown in metal-framed cases.

At the Cathedral of San Lorenzo, in the basement of the Archbishop's Palace, Albini designed the new Tesoro 'chamber' contiguous to the apse and sacristy (Fig. 9.10). Two entrances, one for the clergy and one for the public, led to the small circular space housing the platter of the Holy Grail; a central hexagonal space gave access to other circular spaces of different diameters. The walls of black stone, artificial lighting directed onto individual items and natural lighting from above penetrating though *oeils-de-boeuf* in the roof and striking the radiating beams provide an exceptional example of Albini's idea: 'the atmosphere must not be motionless, stagnant, but must vibrate so that onlookers feel immersed and stimulated without being aware of it'.[59] Albini combined his elements with the surrounding space for maximum effect, balancing the influence between ancient and modern, aiming at communication with a wide public. The museum was considered as part of an urban context perceived by the visitor with a keener appreciation of its historic past.

Conclusions: heritage and the present day

In drawing his conclusions about 'Fifteen years of Italian architecture', Rogers affirmed that the best, most combative Italian architecture had served to 'unfreeze the modern style', to extend the concept of function, to recover a sense of history, but that, even so:

> architecture has not been able to break through the confines of the elite, has not penetrated our society. Exceptions to this are islands in a sea of speculation and mummification: the economic boom, seen in the rising number of new buildings, is not reflected in a diffusion of architectural culture. Proof of this unhappy state is the disfigurement of our country symbolized in our towns and cities: ever less expressive, blissfully unaware, one alike another, each with its own tall building, each 'pulling a face', each with a nervous tic.[60]

The important projects discussed here are representative of the research conducted by a small group of architects who realized that certain principles of the Modern Movement had become impracticable when faced with the realities of post-war reconstruction.

Recently, initial attempts to rehouse the people of l'Aquila, one of the most important cities of art, seriously damaged by the earthquake of April 2009 – and since abandoned – have shown that the problem of reconstruction is not only utilitarian or scientific but a problem of architecture, where regulations and technology need to be combined with research on architectural types and forms.

For these reasons the experience of post-war reconstruction, shorn of its conventional aspects, is still significant today. It is a subject on which contemporary architects and town planners may well reflect: there is a need to appreciate the real character of a town, whatever its size; an untiring search for possible alternatives; responsibility for making decisions that closely affect the users of the architecture; and avoidance of an attitude all too often favourable to higher land values in areas required for other purposes.

Notes

1. *La ricostruzione in Europa nel secondo dopoguerra,* theme issue, *Rassegna,* 54, 1993.
2. Gregotti, V. (1993) 'Editoriale', *Rassegna,* 54: 5.
3. Olmo, C. (1993) 'Temi e realtà della ricostruzione', *Rassegna,* 54: 10.
4. See Samonà, G. (1959/1973) *L'urbanistica e l'avvenire della città,* Bari: Laterza, 203–287; Fabbri, M. (1975) *Le ideologie degli urbanisti nel dopoguerra,* Bari: De Donato; Samonà, G. (1978) *L'unità architettura-urbanistica,* Milan: Franco Angeli; a 1980 issue of *Hinterland,* 13–14, in particular Canella, G. (1980) 'Figura e funzione nell'architettura italiana dal dopoguerra agli anni sessanta', pp. 48–77; Romano, M. (1982) *L'urbanistica in Italia nel periodo dello sviluppo 1942–1980,* Venice: Marsilio; Brunetti, F. (1986) *L'architettura in Italia negli anni della ricostruzione,* Florence: Alinea.
5. Menghini, A.B. (2002) 'The city as form and structure: the urban projects in Italy from the 1920s to the 1980s', *Urban Morphology,* 2: 75–86. Rather than a departure from the models of Rationalism, Italian architecture was able to maintain a unitary culture avoiding division of roles and scales (planning, analysis, design, etc.).
6. Piccinato, G. (2010) 'A brief history of Italian town planning after 1945', *Town Planning Review,* 81: 237–259.
7. In post-war Italy those engaged in architecture and town planning were either qualified engineers or architects.
8. Between 1946 and 1956, almost 1,500,000 people moved from the south of Italy to the large cities of the north, especially Turin and Milan.
9. Italian Neorealist architecture was strongly influenced by the rediscovery of traits that marked Italy's historic townscapes and sought to shape an environment close to the living traditions of the poor. Examples include the 'La Martella' housing scheme at Matera (I. Gardella, A. Quaroni, 1949), the Tiburtino housing district in Rome (L. Quaroni, M. Ridolfi, 1949–54) and the Cesate housing district on the outskirts of Milan (F. Albini, G. Albricci, BBPR, I. Gardella, 1951).
10. Chiodi (1885–1969) was a liberal trained at the Milan Polytechnic, where he began his career in the early 1920s. In the same period he worked as a consultant to the Milan City Council, becoming a key figure in the debate on town planning in the interwar period. He sat on the national commission that drew up the Town Planning Act of August 1942.
11. After studying in Vienna with Adolf Loos, De Finetti (1892–1952) was one of the few Italian architects who kept a European profile even during the Fascist period. A 'loner', ever clear-sighted and realistic, in the interwar period he was among the few Italian intellectuals who believed that survival after Fascism depended on adopting a bourgeois ideology rooted in the European context.
12. Trained at the Milan Polytechnic, in 1939 Secchi (1899–1992) was appointed head of the town planning section in the Engineering and Design department of the Milan City Council. In 1946 he sat on the Central Committee for preparing the Master Plan. Secchi was also famous for rebuilding the Teatro alla Scala, destroyed by the 1943 bombardment.
13. A member of the BBPR group, Rogers (1909–69) was a key figure of the post-war architectural debate. Posing problems of how to preserve existing features of a given urban environment, he explained that modern architecture should be able to give fresh meaning and a new set of values to historic townscapes, putting the principles of the Modern Movement to the test.
14. Astengo (1915–90) pursued research relating town planning to regional planning, socio-economic analysis, and town planning legislation. After 1949 he worked with Adriano Olivetti (one of the promoters of modern architecture in Italy) on refounding the Italian Town Planning Institute (INU), and resuming the publication of *Urbanistica,* in which Italian and international town planning experiences were reported and discussed.
15. Albini (1905–77) was a protagonist of Italian Rationalism whose approach was a personal one embodying elements of traditional architecture. His works included furniture and industrial design. From 1949 he taught at the Venice Institute of Architecture (IUAV), the Turin Polytechnic and the Faculty of Architecture at the Milan Polytechnic.

16. *Metron* appeared in August 1945, edited by L. Piccinato and M. Ridolfi; *Domus,* publication of which continued under Fascism, was edited by E.N. Rogers for 21 issues (from 205 in 1946 to 223–225 in 1947); *Costruzione-Casabella,* edited by F. Albini, appeared at the end of 1946 with two issues devoted to the AR (Reunited Architects) plan for Milan (194) and to Giuseppe Pagano (195–198). Short-lived publications included *A – Cultura della vita,* edited by L. Bò, C. Pagani and B. Zevi (1946–47); *La Nuova Città,* edited by G. Michelucci (1946); and *La Città, Architettura e politica,* edited by G. De Finetti (1945–46).
17. Baffa, M., Morandi, C., Protasoni S. and Rossari, A. (1995) *Il Movimento Studi per l'Architettura 1945–1961,* Bari: Laterza.
18. Zevi, B. (1993) *Zevi su Zevi, Architettura come profezia,* Venice: Marsilio, pp. 52–63.
19. 'Fondazione del gruppo di architetti torinesi "Giuseppe Pagano"' (1945), *Agorà,* 3: 16–18.
20. *Rassegna del primo convegno nazionale per la ricostruzione edilizia: Milano 14–15–16 dicembre 1945,* Milan: Edizioni per la casa, 1945.
21. *Urbanisme et Habitation 1947,* Paris: Edition du Commissariat Général de l'Exposition Internationale de l'Urbanisme et Habitation, 1947.
22. This became law in October 1951.
23. The 1942 Town Planning Act had embodied the fundamental principles of modern town planning. In fulfilment of Article 18, expropriation of land included in the extended areas covered by the plan enabled local administrations to amass vast expanses of land usable for building.
24. See the special issue of *Urbanistica,* 15–16, 1955.
25. Mioni, A. (1979) 'L'urbanistica milanese nella ricostruzione: uomini e strutture', in Bonvini, G. and Scalpelli, A. (eds) *Milano fra guerra e dopoguerra,* Bari: De Donato, p. 550.
26. The National Liberation Committee (CLN) was formed on 9 September 1943. It included the Communist Party, the United Proletarian Socialist Party, the Action Party, the Christian Democratic Party, the Party for Labour Democracy and the Liberal Party. The CLN was deprived of all its functions and dissolved prior to the 1948 elections.
27. The authors of the AR plan were F. Albini, the BBPR group, P. Bottoni, E. Cerutti, I. Gardella, G. Mucchi, M. Pucci and A. Putelli. See *Costruzioni-Casabella,* 194, 1946.
28. Mucchi, G. (1945) 'Studi per il piano regolatore di Milano', *Rinascita,* 11: 250–252, quote from p. 250.
29. The centre of economic activities was to be moved close to the intersection of the highways, adjacent to the Trade Fair.
30. The new Plan for Milan – the first to be approved under the 1942 Act – was the work of a team of technicians coordinated by a central committee and including over 100 professionals forming parallel technical and consulting committees. Among the consultants was Piero Bottoni, who strongly supported a 'regional vision' for the plan. See Bottoni, P. (1955) 'I concetti fondamentali del nuovo Piano regolatore', *Urbanistica,* 15–16: 197–201.
31. Chiodi, C. (1944) *Studi e proposte degli ingegneri milanesi intorno ai problemi edilizi della ricostruzione della città,* Milan: Officina Grafica Suppi.
32. Morandi, C. (1999) 'Urbanista e *civil servant* della città di Milano', in Susani, E. (ed.) *Milano dietro le quinte, Luigi Lorenzo Secchi,* Milan: Electa, p. 122.
33. See Secchi, L.L. (1957) 'Il progetto del "p.r. 1945" presupposto al nuovo piano regolatore generale della città di Milano', in *Aspetti problemi realizzazioni di Milano. Raccolta di scritti in onore di Cesare Chiodi,* Milan: A. Giuffrè, pp. 513–525.
34. Ibid., p. 521.
35. De Finetti, G. (1946/1969) 'Su un concorso di idee per rifare Milano: aspetti spirituali ed aspetti tecnici della cosa', *La Città,* 2, reprinted in De Finetti, G. *Milano. Costruzione di una città,* Milan: Etas Kompass, pp. 450–454.
36. *Milano Risorge* is the title of a book that De Finetti wrote but never published. It is included in De Finetti, *Milano. Costruzione di una città,* together with studies and projects that were unpublished, or published in newspapers and journals.
37. De Finetti, G. (1946/1969) *Considerazioni sull'edificio sorgente in Milano, piazza del Duomo-corso Vittorio Emanuele-via Agnello,* draft copy, reprinted in De Finetti, *Milano. Costruzione di una città,* pp. 397–399.
38. The main streets in Roman military camps, the *cardo* (N-S) and the *decumanus maximus* (E-W), were also a basic feature of Roman city planning, the Forum normally being located at the intersection of the two streets.
39. In 1938 the 14 Fiat factories employed 45,000 workers and 5,000 office staff; by 1941 the population of Turin was double what it had been in the early twentieth century.

40. Astengo, G., Bianco, M., Renacco, N. and Rizzotti A. (1947) 'Piano Regionale Piemontese', *Metron,* 14: 3–29.
41. Astengo, G., Renacco, N. and Rizzotti, A. (1953) 'La pianificazione delle regioni italiane. Piemonte', in *La pianificazione regionale,* Rome: INU, p. 387.
42. Scrivano, P. (2000) 'The elusive polemics of theory and practice: Giovanni Astengo, Giorgio Rigotti and post-war debate over the plan for Turin', *Planning Perspectives,* 15: 3–24.
43. The Fiat Mirafiori factory was built in 1937 for 22,000 workers.
44. Astengo et al., 'La pianificazione delle regioni italiane. Piemonte', make explicit reference to a number of sources.
45. Ibid., figure 9.6.
46. Though seemingly modern, it was in fact deeply rooted in the pre-war 'paternalistic' idea of building working-class villages next to the industries where the inhabitants would work.
47. G. Astengo designed the Quartiere Falchera with S. Molli-Boffa, M. Passanti, N. Renacco and A. Rizzotti. See Astengo, G. (1954) 'Falchera', *Metron,* 53–54: 13–63.
48. In February 1944 Olivetti paid a secret visit to Switzerland, where he worked on his plan for reforming the State. See Olivetti, A. (1946) *L'ordine politico della Comunità. Dello stato secondo le leggi dello spirito,* Rome: Edizioni di Comunità. Olivetti put his ideas into practice at Ivrea, his democratic factory-city. See Pampaloni, G. (1974) *Architettura e urbanistica negli Anni Cinquanta alla Olivetti,* Florence: Officine Grafiche.
49. INA-Casa (1949) *Suggerimenti, norme e schemi per l'elaborazione e presentazione dei progetti,* Rome: Tipografia F. Damasso.
50. INA-Casa (1951) *Suggerimenti, esempi e norme per la progettazione urbanistica,* Rome: Tipografia M. Danesi.
51. *Urbanisme et Habitation 1947,* p. 42.
52. Rogers, E.N. (1958/1997) 'Il Cuore: problema umano della città', reprinted in Rogers, E.N. *Esperienza dell'architettura,* Milan: Skira, pp. 257–260; authors' own translation.
53. Rogers, E.N. (1954) 'Le responsabilità verso la tradizione', *Casabella-Continuità,* 202: 1–3; (1955) 'Le preesistenze ambientali e i temi pratici contemporanei', *Casabella-Continuità,* 204: 3–6; (1955) 'La tradizione dell'architettura moderna italiana', *Casabella-Continuità,* 206: 2–6; (1956) 'L'architettura moderna dopo la generazione dei Maestri', *Casabella-Continuità,* 211: 1–5; (1957) 'Continuità o crisi?', *Casabella-Continuità,* 215, p. 1.
54. Argan, G.C. (1949) 'Il museo come scuola', *Comunità,* 3, p. 64.
55. Albini, F. (2006) 'Le mie esperienze di architetto nelle esposizioni in Italia e all'estero', lecture given at Venice University at the opening of the 1954–55 academic year, republished in Bucci, F., Irace, F. (eds) *Zero Gravity Franco Albini, Costruire la modernità,* Milan: Electa, pp. 75–77, quote from p. 76.
56. Marcenaro (1906–76) had taken part in the Resistance movement in Genoa and in 1950 was appointed Director of Antiquities, Fine Arts and History.
57. Bordered by 11 large palaces between 1558 and 1583, this 250m street was later completed by the Palazzo Rosso (1671) and Palazzo Bianco (1714).
58. Emiliani, M.D. (1982) 'Musei della ricostruzione in Italia, tra disfatta e rivincita della storia', in Magagnato, L. (ed.) *Carlo Scarpa a Castelvecchio,* Milan: Edizioni di Comunità, pp. 149–170.
59. Albini, 'Le mie esperienze di architetto nelle esposizioni in Italia e all'estero', pp. 75–77, quote from p. 76.
60. Rogers, E.N. (1961) 'Il passo da fare', *Casabella-Continuità,* 261: 3.

10

RHETORICS AND POLITICS

Polish architectural modernism in the early post-war years

David I. Snyder

Following the Second World War, the reconstruction of European cities was a principal concern of architects, planners, political authorities, and private citizens alike.[1] Between the liberation of Polish territories by the Red Army in May 1945 and the final consolidation of the Soviet-backed post-war socialist government in December 1948,[2] recovery from the physical devastation wrought by war in Poland went far beyond the pragmatic concerns of rebuilding ruined cities and providing shelter for displaced populations. Over the course of three years characterized by political uncertainty and a state of virtual civil war in Poland, reconstruction plans were developed for Warsaw that, in tandem with the incremental ascent of the Soviet-backed socialist government, subtly infused the projected image of the rebuilt icon of national resistance and martyrdom with new, ideologically driven meanings. While many features of its theoretical basis and tactics for implementation were not entirely unique to the Polish context, a distinctive planning strategy and reconstruction methodology nonetheless emerged that ultimately provided the animating life force for a rebuilt Warsaw as the indisputable emblem of post-war Polish national identity,[3] one equally shaped by political ideology and by spectatorial considerations derived from the photographic image. Revealed in June 1949 as the *Six-Year Plan for the Reconstruction of Warsaw*,[4] the final scheme for rebuilding Warsaw incorporated a radical shift in the relation between the spatial and temporal domains wholly predicated on a synthesis of ideology and ways of viewing and attaching meaning to the built landscape of the Polish capital.

For two months, in May and June 1945, the *Warsaw Accuses*[5] exhibition occupied the galleries of the National Museum in Warsaw before it was sent across the Atlantic Ocean as a travelling exhibition in the United States (Figs 10.1 and 10.2). As stated clearly in the official exhibition catalogue, its curatorial agenda was to memorialize the violence endured by Warsaw and sanctify its memory by binding the image of its destruction to the vision of a more beautiful future. Curiously, while the city lay in ruins visible to all, artefacts were retrieved from the rubble and put on display alongside the photographic record of the city's undamaged pre-war state. One result was that the before-and-after photographic juxtapositions inside the museum performed as the putatively complete representation of a fragmented and imperfect external reality. By directing the attention of the viewer to the image of destruction rather than to the actual physical ruins of the city, the format of the exhibition helped to set in motion a process in which the architectural object's relationship to historical time and perceivable space became irrevocably altered. Essentially, the internal temporal sequencing of the exhibition derived from the compound images of pre-war wholeness and post-war destruction was intended to provide the Polish audience with a visual index for measuring the extent of destruction inflicted on Warsaw. However, the spectatorial procedures inside the museum established how photography is perceived and operates in common perception in a way that fundamentally shifted the derivation of meaning from, and its attachment to, the urban context. In other words, in the reconstruction of Warsaw, photographs (technically reproducible images), as opposed to images held in the memory (subjective cognitive snapshots), to paraphrase Siegfried Kracauer, 'perform a mediating function' that annihilates the portrayed subject.[6] In re-examining the reconstruction of Warsaw with the aid of Kracauer's theoretical lens, a string of pivotal, fundamentally interrelated questions arises: to what extent does this reliance on images compound the fragmentation of urban space, detach social experience from the material and spatial context, eviscerate or augment memory, and contribute to the spectacularization of the city? How is the sense of place determined by representational strategies rooted in the nomenclature of photography? And, more broadly, what are some of the implications for larger conceptions of architectural modernism that can be gleaned from the experience in post-war Poland?

The use of multiple photographic pairings in post-war Poland ostensibly strove to become what Kracauer defines, in his 1927 essay entitled 'Photography', as memory images. While he argues that the photograph and the memory image are essentially different and fundamentally at odds with each other, the paradigm established in the *Warsaw Accuses* exhibition aspired to invest the images on display with the kind of meaning he considers unattainable by photography. Precisely because the architectural object portrayed was inextricably linked to the image of its ruin, the photographic pairings reify rather than banish the recollection of death, which Kracauer claims to be 'part and parcel of every memory image'.[7] Moreover, they acted to simultaneously disengage the architectural object from both its physical (spatial) and historical (temporal) contexts, again in opposition to his assessment of the essential qualities of the photograph. Memorializing the buildings and urban spaces of Warsaw as victims through photographic representation in the end became a benchmark for measuring the cultural-historical value of any given building or historical streetscape. In turn, this new, recalibrated form of memory image, no longer confined to the lived experience of the individual subject, provided the inspiration for the projected material future of post-war Warsaw and its place within the national consciousness. A 'properly reconstructed' Warsaw in which historic landmarks would stand alongside new construction, therefore, had to contain architectural elements that somehow would be construed by the broader Polish audience as reflecting Polish national identity, while also satisfying the ideological constraints of the emerging socialist political regime. The synthesis between nationalist and socialist agendas – by definition antagonistic ideologies – was a key component of the socio-political evolution of post-war Poland and found fertile ground for its implementation and dissemination in the reconstruction of Warsaw. All that was required was a public schooled in decoding the symbolic content of the city.

In developing protocols for inculcating the masses with the necessary perceptual apparatus for assimilating this revamped message of national identification embedded in the city, architects, planners, and bureaucrats seized upon the persuasive emotional power of the photographic image. They deployed it as a didactic tool that would create uniform memories of the

FIGURE 10.1 Royal Castle, Warsaw. Pre-war (above), 1945 (below).

FIGURE 10.2 Bruehl Palace, Warsaw. Pre-war (above), 1945 (below).

city as it had been and codify the ways in which the symbolic content of the reconstructed city would be understood in the future. The centrepiece of the call to rebuild Warsaw was the symbolic heart of the city, the Old Town (Stare Miasto), insofar as this quarter constituted the primary site in the nascent socialist metropolis where Polish resistance and heroism, victimization and martyrdom, and the resurrection of the Polish national spirit and restoration of Poland's cultural heritage could be synthesized. Whether or not conceived as such from its inception, the scheme for rebuilding Warsaw's historic medieval urban core became the model for shaping public opinion about reconstruction methods and forging uniform collective memory in post-war Poland. And just a few months after the mounting of the *Warsaw Accuses* exhibition, the illustrated popular press was enlisted as the ideal vehicle for disseminating the reconstruction plan and convincing the broadest segment of the Polish audience of its inherent truths.

Publicity and identity

The nationally distributed illustrated magazine *Stolica* (Capital city) was an official publication of the Bureau for the Reconstruction of the Capital (BOS), the government agency in charge of rebuilding Warsaw. Articles, reports on cultural activities, and editorial commentary published in *Stolica* kept the public informed of all matters connected to reconstruction, and it became, in effect, a weekly progress report issued by BOS for the Polish lay audience. Overall, this illustrated magazine delivered a body of information whose message oscillated between promoting the overarching ideological aims of the state and asserting the authenticity of the rebuilt architectural markers of national identity. Primarily using images and accompanying texts to provide a historical chronicle of the city, the editors of *Stolica* established a series of new associations around architecture that substantiated the continuity between purported historical facts and current realities. Specifically, they relied on the durability of images once they had entered the public consciousness and their subsequent associational flexibility to impart the magazine's central message and legitimize the regime's credo: that the 'New Socialist Warsaw' rising from the ashes was indeed the fulfilment of Poland's national destiny.

The main article in the introductory issue of *Stolica,* from 3 November 1946, was a photo-essay entitled: 'Beautiful Warsaw that is, Alas, No More, and which We Will Resurrect'.[8] In it, the core components of the reconstruction are identified, and all the ideological arguments in support of rebuilding particular historic monuments are given. Employing the language of national and historical identity, the article drove home the argument that reconstruction was part of a divine historical plan for the nation by tying the rhetoric of the text to the familiar optical device of before-and-after images.

> Warsaw has her own everlasting beauty. She has not lost it even now when many of her most beautiful monuments have suffered total annihilation. There is no dispute, there are no two opposing opinions in the heart of the Polish community, that the recognized architectural monuments and cultural property of Warsaw, from the Royal Castle to the Cathedral to the Old Town must be resurrected.[9]

Further on, with surprising candour, the editors of *Stolica* acknowledge the inherent disjunction between a reconstructed Old Town and its pre-war authenticity as the product of lived human experience.

> It has been said, among other things, even if it succeeds as a faithful reproduction and the old monumental architectonic form is given to the ruined historical monuments that it will not be the same as it was before. Because that which lasted half a century or 500 years, that which absorbed in its own walls the experiences, thoughts, suffering of every generation, that cannot be restored through resurrection. That is irreplaceable, like the life of a human being.[10]

On one level, the *Stolica* editors' argument acknowledges the irreplaceable aspect of this tremendous loss. But is there another, more potent essence to their highly charged emotional argument connected to the particular methods they employed to assist the Polish public to overcome irreplaceable losses, both human and material, in the wake of great tragedy? Indeed, how the editors of *Stolica* resolved this dilemma confirms the utility of images in developing methodologies for re-educating the masses. It demonstrates how publicity and photographs of architecture were systematically and inextricably bound to the image of the reconstituted future.

> However, certainly just as a picture of the deceased is pleasing to those who are left behind, although it is something distinct from the person himself who is pictured, that same faithful architectural copy of the most precious monuments – the Royal Castle, the Cathedral, the Old Town Square – will be a route for all residents of Warsaw and every Pole; like a memento, that which is irrevocably lost to the past will be established as the object of national pride. He who *looks* [emphasis added] today at pre-war photographs, awakens in himself a noble pride. . . .[11]

Architecture in its reconstructed form is likened to a photograph, and this conceptual armature that positioned the reconstructed city as an image of itself overtly defined the spectatorial agenda of the emerging socialist metropolis. Moreover, the article claims, through the act of viewing the old photographs and by extension, the rebuilt historic landmarks,

> that [former] beauty will be resurrected. At first it may certainly be in small quantities and perhaps superficial, but with time the new walls will become imbued with the rejuvenated atmosphere of the Old Town, an atmosphere which constantly dwells there.[12]

The claim here is that the act of viewing old photos in and of itself becomes the conduit for imparting the message of authenticity to the public. According to this equation, images of the old validated the materiality of the new socialist metropolis rising in the landscape and to some degree restored Warsaw's pre-war essence. Moreover, the explicit use of the Polish verb for 'resurrection' (*wskrzesić*) in the title rather than 'rebuilding' or 'reconstruction' demonstrates how the editorial staff played on the emotions of the public. For the predominantly Roman Catholic audience, this particular verb had an incontrovertible resonance; it articulated the martyr status of Warsaw and reinforced the well-established Polish self-image as 'the Christ of nations' (*Polska Chrystusem narodów*),[13] thereby emphasizing the sense that reconstruction was part of a divine historical plan for the nation.

Subsequently, and with increasing intensity, the present was delivered to the public on the pages of *Stolica* as being fundamentally contingent on a particular image of the past held in the memory, which in turn was translated into an inspiration for the reconstituted future. Images from the 'New Warsaw' that began to appear in magazines like *Stolica* brought the successes of rebuilding Warsaw and the ingredients for shaping a new national identity into every home in Poland, obviating the need to see Warsaw at first hand as it was being rebuilt. The rebuilding campaign for Warsaw echoed the reconstruction of Polish national identity through a synthesis of two key ingredients: first, through a revised historical narrative and, second, by means of an ideologically predetermined prognosis for the future that was derived from the logic of the before-and-after photographic model. To borrow a theme developed by Katerina Clark, a specialist in Soviet-era literature, this temporal and spatial sequence followed the rule of 'modal schizophrenia'. It collapsed the distance between 'what is' and 'what ought to be', a condition she asserts was a primary operational tactic in all post-war socialist societies.[14]

Thus, as the reconstruction of Warsaw progressed, the city as a collection of ruins emerged as the 'before' image, slowly supplanting the catalogue of pre-war photographs. The reworked historical reconstructions and the triumphant new socialist spaces became the new 'after' images, where the success of the socialist project was verified. In the end, the potential ambiguity of elements from the old urban landscape within the new socialist metropolis was displaced by the historical weight and authority attached to the reconstructed monuments of Warsaw. This relationship between the old and new in the socialist city was 'as much about managing the meanings and associations of the historic city as it was about the modernisation of the urban fabric', according to the cultural historian David Crowley.[15] The use of photographic images in post-war Warsaw, their assimilation by the viewer, and ultimately their inscription on the rebuilt city played an essential role in this 'management of meanings'; it demonstrates the particular relationship between history and modernity forged by architecture in the emerging socialist reality. Revision of ideologically objectionable architecture, either through total erasure or through 'scientific' modification (meaning preservation methodologies), was justified as a more complete and accurate process of restoration. Moreover, with the images of Warsaw's destruction now constituting the datum against which the new socialist architecture was to be measured, the fullness of the city's history was radically abbreviated. Thus, in shaping the plan for the 'New Socialist Warsaw' and purging it of any ambiguous signs of a more differentiated Polish past, the re-rendered history of the city and the nation as constituted by architectural monuments effectively lost all specificity. In theory, the resulting urban mélange of old and new predicated on the memory of destruction and contextualized within the ideologically saturated socialist metropolis would impart to every visitor to Warsaw a clear sense of the new Polish identity, an identity firmly grounded on allegedly authentic and irrefutable historical cues.

The past-present/old-new dichotomy was codified and commodified as the common property of every citizen in the People's Republic of Poland. In the pages of publications such as *Stolica* the 'old' and 'new' photographs provided the foundation for the new and uniform collective identity. The value of images as ciphers to some higher spiritual truth, as argued by the cultural historian David Gross, is due to their volatility and instability insofar as 'they carry no precise content or message which can be readily agreed upon, but are instead open-ended and continuously reinscribable with new meanings to replace old ones'.[16] It is precisely the malleability of the photograph, as a marker of some higher meaning, which the editors of *Stolica* relied upon to broadcast their message to the widest audience. Subsequently, the formula linking the past to the future was ensconced permanently in the exhibition space of the Museum of the City of Warsaw, located on the Old Town Square.

The City Museum opened its doors to great fanfare in 1953, coinciding with the official dedication of the rebuilt Stare Miasto.[17] Upon its establishment, a new phase in the instrumentality of photographic images was inaugurated. In the Old Town, the essential connection between the modern city and the revamped historical narrative of the nation was made manifest. This happened in two ways: first by establishing a correspondence between the museum space, where the narrative was laid out and imparted to the visitor; and subsequently in the recently completed urban setting, where each individual actively participated in the enactment of this story of national suffering and redemption simply by seeing the end results. The visitor to the Museum of the City of Warsaw, who it was assumed was already well-versed in the procedures for decoding the symbolic content of photographic images after years of flipping through the pages of *Stolica,* would advance her education in the exhibition space of the museum. Meandering across time and through space and ending up in the galleries where dioramas documented the destruction of the war years, the spectator upon exiting the museum would then arrive at a material present that functioned as the fulfilment of the ideologically reworked history of the nation – namely, the complete and putatively authentically rebuilt Old Town (Figs 10.3, 10.4). This final phase of development in the image-based architectural production characteristic of post-war Poland now promoted a reading of the Old Town as the three-dimensional image of the larger historical city, but with a new twist.

The content and disposition of the before-and-after images were modified yet again, so that the 'before' image had now become the photographic display inside the museum while the new 'after' image was the full-scale architectural representation of 'Old Warsaw' that was instantiated in the rebuilt environment. The evolution of this architectural promenade took the visual cues learned by the visitor inside the museum and then not only implemented them on the streets of the rebuilt city, but also placed the spectator inside the frame of the image. Curiously, this constituted a complete inversion of the paradigm established eight years earlier in 1945 in the *Warsaw Accuses* exhibition. Then, the attention of the spectator was directed to the interior of the museum, towards the photographic pairings and away from the actual ruins of the city. By 1953, the spectator's gaze was directed towards the material spectacle of the reconstructed city: first in the Old Town, where the inscription of the historical nation was located in the rebuilt historical structures, and then in the new metropolis, where the Polish nation's socialist destiny was actualized and its post-war identity verified. Insofar as the representational strategies that determined the presentation of history inside the City Museum were reproduced in the rebuilt spaces of both the historic Old Town and the new socialist Warsaw beyond, and in the visual register established by images of Warsaw in ruins coupled with the present-day photographs of a reconstructed city, the architectural object was liberated from its historical context and all the attendant social, economic, and cultural influences. The Old Town was not a 'faithful replica' of its pre-war condition but instead a collection of symbolic representations that signified an authentic and ideologically amenable form of Polishness. Similarly, the newly constructed areas of the city and the heroic spaces in which the theatre of the socialist masses was choreographed and ritualistically performed were also determined by comparison with their post-war ruinous state. Thus, in the spatial and temporal logic of the *Six-Year Plan for the Reconstruction of Warsaw,* 1945 became the true starting point of Polish history. It was the year when a new visual language and rigidly defined collective memory were integrated into the architectural culture of post-war Poland. No longer contingent on pre-war photographic snapshots documenting the historical visual aspect of the city, the architectural objects and urban spaces of post-war Warsaw, both old and new, were cleansed of all ambiguity and multiple meanings and transformed into homogeneous affirmations of Polish national destiny.

FIGURE 10.3 Old Town Castle Square, Warsaw. 1945 (above), 1953 (below).

FIGURE 10.4 Old Town Market Square, Warsaw. 1945 (above), 1953 (below).

In retrospect, the introductory passage from the *Warsaw Accuses* exhibition catalogue provides an unambiguous key for decoding the driving forces behind the post-war reconstruction of Warsaw.

> The spectator, entering the newly-opened Museum in Warsaw, in particular if he is by birth a child of this martyr-city, will be – as we think – doubly moved and excited. The building itself of the institution, raised very rapidly from degradation and ruin, which it was driven into by German violence, is a real token foretelling the rise of a new, more beautiful capital. This museum not only suffered the most painful losses of all Polish Museums, but had to look at the whole martyrdom of the old culture of the capital. We witness a strange community of martyrdom, enchanting in its pathos, between the destinies of living Polish beings and of dead objects, the material exponents of Polish culture. A new generation will grow up, the walls of Old Warsaw will rise from ruins.[18]

At the core of the catalogue's message lies the notion that Warsaw, having suffered the devastation of German wartime aggression directed specifically towards the eradication of Polish culture, was a 'martyr-city' whose restoration and reconstruction would be the fulfilment of Polish national destiny. A mere three months after Poland's liberation, rebuilding Warsaw was articulated here as both an instinctive reflex and a moral obligation. Moreover, anchored in the perception of the city as the quintessential material reflection of Polish identity, reconstruction entailed recovering Warsaw's particular sense of place out of the ruins. This sanctification of the built landscape went much deeper than mourning or nostalgic yearning for what had been lost in the war. But, in May 1945, amongst architects and planners, exactly what had been lost and how it was going to be retrieved remained unanswered.

The architectural discourse

In February 1946, just eight months after it closed at the National Museum in Warsaw, the *Warsaw Accuses* exhibition was sent to America accompanied by a delegation of architects from BOS, including Stanisław Albrecht, Aleksy Czerwiński, Tadeusz Głogowski, Helena Syrkus, and Szymon Syrkus.[19] The *Warsaw Accuses* exhibition opened at the Library of Congress on 16 April 1946, timed to coincide with the annual Washington Chapter convention of the American Institute of Architects. Walter Gropius delivered the opening remarks, and Szymon Syrkus gave the keynote address.[20] Gropius ceremoniously inaugurated the opening of the exhibition by stating:

> For a few years now, I have been a citizen of the United States of America. I left Germany in 1934 against my will because Nazism was an abomination to me. But I am, nonetheless, a German from birth and because of that, I share some responsibility for the acts perpetrated against Warsaw by the Nazis. I am full of admiration for Polish architects and urbanists, many of whom I count among my closest friends, who despite the threat of mortal danger continued to work on plans for the building and reconstruction of their national capital throughout the entire occupation, which they have translated into a planning strategy for a new reality, based on creative principles that stem directly from the CIAM congresses.[21]

By linking the post-war reconstruction plans for Warsaw to pre-war strategies developed under the aegis of Polish architects affiliated with CIAM (Congrès Internationaux d'Architecture Moderne), Gropius was asserting the existence of both an ideological and a practical continuity between the pre-war international modernist agenda and post-war planning and reconstruction methods in the Polish context. Szymon Syrkus, the leader of the Polish delegation, spoke to the audience about the daunting task of reconstruction, which not only had to provide shelter for hundreds of thousands of displaced and homeless Poles and restore Warsaw's devastated infrastructure but also had to balance immediate needs with a master plan that addressed the emerging economic, social, political, and cultural reality. With the vocal support of Gropius and the tacit endorsement of the Polish interim government, the first version of the reconstruction plan for Warsaw was presented publicly. Not surprisingly, this was a scheme based largely on the Functional Warsaw plan created by Syrkus and his colleague Jan Chmielewski for the 1933 CIAM IV Congress in Athens.[22] Even though the realities of war had irrevocably and radically altered the material landscape of the Polish capital, in 1946 as Poland's political future remained uncertain, both Gropius and Syrkus were making the claim that the continuity of pre-war and post-war Polish modernism had not yet been disrupted.[23]

Adding to the bittersweet tone of this auspicious occasion, Lewis Mumford, who had prepared the introduction for the American exhibition catalogue, suggested that its title be altered to *Warsaw Accuses – Warsaw Lives.*[24] In comments that appeared the following week in the *New York Times,* Mumford characterized the relationship between the photographic record of Warsaw's destruction and the post-war reconstruction plan as 'engaged in fitting these remnants of the past into a new city which is to meet the needs of a machine age'.[25] One year after Poland's liberation and in the symbolic heart of Western democracy, a two-pronged programmatic basis for Warsaw's reconstruction as yet unaffected by socialist realist ideology was unveiled, in which the restoration of architectural monuments and historical districts constituted an indispensable component of an otherwise thoroughly functionalist modern master plan. Photography served as the binding thread linking the two vectors of reconstruction and bridging the gap between the past and the future.

Complementing the travelling exhibition and lecture circuit of prominent Polish architects across the United States, in February 1946 the Polish Embassy began publishing a monthly bulletin in English entitled *Poland of Today.* Targeted at the American public, it aimed to spark sympathy for Poland's predicament, encourage economic support for reconstruction efforts, and, perhaps most important, portray the post-war People's Democracy of Poland as a nation shaped by the same principles that gave rise to America. In addition to underscoring equality and pluralism as the basis for post-war Polish society, the rhetoric of democracy was a means of dissipating fear and American anxiety over communism and the growing Soviet sphere of influence in Eastern Europe. Like numerous other publications in Poland that brought the ideological message of the post-war regime to the public, using pictures and fact-filled essays to report on reconstruction efforts, *Poland of Today* was an effective tool geared towards reshaping American perceptions of Poland and the emerging Polish government. Not surprisingly, articles graced by heroic titles such as 'Poland Will Not Succumb to Despair' or 'Poland's Museums and Libraries Desecrated', and the more straightforward 'Reconstruction of Warsaw', were the English-language counterparts to similar articles written for the home audience in Polish periodicals. Mobilizing public opinion had both national and international objectives.

Relying on the emotional impact of the haunting images of Warsaw in ruins, authors turned to catch-phrases like the 'spontaneous repopulation of Warsaw leading to a new city rising out of the ruins', and 'today there are no big capitalists or estate owners in Warsaw; they have ceased to exist in Poland', to subtly disseminate the underlying social-democratic political message.[26] Although the consolidation of the government had not yet been achieved, by 1946 the anti-capitalist, broadly based socialist foundation of post-war Polish political reality was firmly in place. 'The main idea of the reconstruction plan is to preserve the city's characteristic outlines, its former architectural profile, as it were, while at the same time avoiding the confusion and overcrowding that in some areas prevailed so widely'.[27] Reshaping the material, spatial, and social contours of the Polish capital was framed as both an emotionally restorative and a politically corrective endeavour. On the one hand, it would right the historical wrongs perpetrated against the Polish people while, on the other hand, validating the vision of Poland's future promoted by the nascent post-war regime. Curiously, not only did socialist ideology influence the direction of architecture and urban planning, but, as the new Warsaw took shape, architecture and planning reciprocated by providing the material evidence that legitimized the central claims of socialism.

Building a new Warsaw was both an act of retrieval and commemoration as well as a forward-looking enterprise which, in theory, was achievable but much more difficult to realize in practice. As revealed in an essay written by the urban planner Adam Kotarbiński in 1945, the essential problem pivoted around one fundamental question: reconstruction or rebuilding?[28] Without providing a direct answer, he nonetheless expresses no doubt that 'the resulting city will be splendid: new – and old'. According to Kotarbiński's diagnosis, the primary task before architects and planners is the acquisition of new knowledge needed to create this desired synthesis. For him, the new Warsaw somehow must incorporate the pathos of the devastation meted out by the Germans and attend to the needs of post-war Polish society. While relieving the congestion typical of city centres, remedying the problems of housing and transportation, and reversing all of Warsaw's former defects, he advises designers to constantly ask themselves, '[W]here does Old Warsaw have to be [in the new scheme]?'[29] Where does the old fit snugly with the new? Although a definitive, ideologically determined answer would not be reached until four years later when socialist realism was officially declared to be the only viable mode of modern architectural expression in a truly socialist society, a workable and highly effective formula for integrating past and present was close at hand, and the key was the before-and-after photographic visual register laid out in the *Warsaw Accuses* exhibition.

Remarkably, at the same time that most architects at BOS were adjusting to an increasingly monolithic, ideologically driven definition of architectural modernism, the political situation in 1947 still allowed for the expression of divergent opinions. In what was no doubt one of his last published endorsements of international architectural modernism, Szymon Syrkus repudiated the nineteenth-century capitalist city and exposed his deep commitment to the modernist principles articulated in his pre-war works, particularly the vision of modern Warsaw he had presented at the CIAM IV Congress in 1933.[30]

> The class-needs of workers are different than the pre-war bourgeoisie. We have no intention of rebuilding the sins of the nineteenth and early twentieth centuries, or to reflect the eagerness and desire for profits that were rampant among the speculators in tenement housing and industrialists.[31]

Modern architecture and the modern metropolis, for Syrkus, were effective instruments of social reform. They were a means of fulfilling the prediction that 'the former Warsaw of the bourgeoisie will be the workers' Warsaw. Workers declare: we do not want an old city, we want a new city – we want a New Warsaw!'[32] Not surprisingly, his comments coincided with the implementation of the first three-year plan, a provisional scheme that provided housing, opened up new industrial enterprises, and rebuilt the infrastructure of Warsaw for the capital city's growing population. Less than a year later, voices like Syrkus' would be silenced and the modern architectural discourse in Poland transformed into a prescriptive monologue.

The penultimate moment in this developmental trajectory of post-war architectural theory came in 1948. Enumerating the major milestones of Polish history, the novelist and poet Julian Wołoszynowski binds their significance to Warsaw and the nation by invoking the ethos of the exhibition, in particular the before-and-after photographic record that the *Warsaw Accuses* exhibition promoted as the authentic and authoritative archive of truth and meaning.

> The history of this city, the capital since the seventeenth century, that was made manifest through the beautiful architecture of the Stanislavian era, the industrial and commercial expansion of the nineteenth century, through two uprisings – November and January, in the last war acquired a symbolic pathos. . . .Ten years ago, in the newly erected National Museum, there was the exhibition, 'Warsaw – Yesterday, Today, Tomorrow'. . . . after the war another exhibition in the same museum, 'Warsaw Accuses' provided an addendum to the previous catalogue.[33]

Wołoszynowski's recalibration of historical time according to the collapsed temporal logic employed in the *Warsaw Accuses* exhibits foreshadowed larger realignments on the political horizon that inevitably led to an ideological shift in post-war Polish architecture. Fusion of political ideology, reconstruction methodologies, the role of historic preservation, and the overarching issues of modern town planning coalesced and operated under a new rubric of materiality, spatial dynamics in the urban context, and temporal sequencing.

When political ideology and architecture intersect

With the appointment of Bolesław Bierut as Secretary General of the Central Committee of the PPR (Polish Worker's Party) in September 1948, the Soviet-backed takeover of Poland was complete.[34] Once political power was consolidated and with the aid of a powerful propaganda machine, Bierut and his supporters were able to exploit the national component of socialist realism in order to translate Polish patriotic sentiment into tangible political gains. More specifically, they harnessed the profound emotional and psychological attachments Poles had to their capital city and the universal commitment to reconstruction through a combination of new building projects, words, and images that conflated socialist ideology with the familiar tropes of heroic defiance against foreign enemies, martyrdom, and the redemption of the nation. Consequently, Warsaw's reconstruction was configured not only as a national project but, more poignantly, as the fulfilment of divine destiny. The image of the new Polish capital that was to be presented a year later in the *Six-Year Plan for the Reconstruction of Warsaw* became the unequivocal symbolic surrogate for national suffering and redemption, and the materialization of the long-imagined utopian Warsaw.

The culmination of this evolution came in the summer of 1949 with the installation of socialist realism as the only viable modern architectural method in the Polish People's

Democracy[35] and the subsequent public release of the *Six-Year Plan for the Reconstruction of Warsaw.* On 20–21 June 1949, at the conference of Party Architects, Edmund Goldzamt, Jan Minorski, and Stefan Tworkowski presented the ideological and methodological foundations of socialist realism to the architectural community.[36] Simply through the 'scientific' engineering of spatial relations (scale and proportion), structures (construction techniques), and forms (stylistic classifications), socialist realism presupposed that if the environmental conditions for modern life in the city were built correctly and in accordance with established theoretical paradigms, then the beliefs and behaviours of its inhabitants would automatically follow suit. Such a theoretically determined position betrays an unswerving faith in the calculability of living conditions, the social and material parameters of the environment, and, most important, the fundamental doctrine of socialism whereby everyday life and human intercourse have no unquantifiable or unexpected contingencies. Not only would the modern socialist metropolis cater to the basic functional needs of its citizens by providing equal access to adequate housing, food, education, employment, cultural entertainment, and leisure activities, but, more important, once its formulaic approach to modern urban living was set in motion, such a system would be self-perpetuating and thus self-validating.

In July 1949, just a few weeks after the era of socialist realism in Poland was officially ushered in at the annual convention of Party Architects, Bierut publicly unveiled the *Six-Year Plan for the Reconstruction of Warsaw.* His opening remarks framed the presentation of the plan by using the language and motifs of pre-war Polish nationalism to describe the most recent national trauma.

> To the Nazi criminals who, breathing hatred and revenge, were, shortly before their own mortal defeat, systematically destroying the city, burning and demolishing for months house after house, in order that no stone of Warsaw might be left upon stone – to them the final destruction of the city seemed an irrevocable fact. As late as January 1945 . . . special squads of destroyers of Warsaw were with sadistic precision boring holes in the walls and remaining historic buildings. . . . Six months of planned destructive activity was not sufficient to raze everything to the ground, as was the case almost two years earlier in the northern part of the city – the Ghetto. All the same, the Nazi criminals left Warsaw convinced that nothing could ever bring about its resurrection. . . . It was clear for the majority of [the inhabitants of Warsaw] what counted above all was their love for the heroic city and their veneration for the sacred ruins of Warsaw.[37]

By infusing the *Six-Year Plan*'s vision of the future with familiar meanings, particularly the suffering and martyrdom of the Polish people, Bierut presented himself as a messianic figure and socialist realism as the new operational dogma for achieving national redemption. The *Six-Year Plan* simultaneously decried the faults and failures of the past while reifying their destruction as the ultimate emblem of national martyrdom. Thus, a new set of photographic pairings depicting Warsaw in ruins as the 'before' image were linked to 'after' images of recently built, modern residential and commercial projects (Figs 10.5, 10.6). The new city that would emerge from the ruins would not only rectify the ills of the past but stand as a shining example of the ascendant Polish nation finally taking control over its capital city and national destiny. Concomitant with the ideological encryption of Warsaw's new material and spatial contours, the modernist discourse in architecture was suppressed.

FIGURE 10.5 Marszałkowska Street near Wilcza Street, Warsaw. 1945 (left), 1953 (right).

FIGURE 10.6 Marszałkowska Street near Świętokrzyska Street, Warsaw. 1945 (left), 1953 (right).

The underlying claim that the city in its urban configuration and architectural appearance would serve as the material embodiment of the post-war nation simultaneously turned to the past by restoring specific historic areas for its substantive basis while looking to the future and declaring the newly planned public spaces and residential districts as the fulfilment of Polish national destiny. This accorded with the theoretical and operational tactics of socialist realism, which, in forging a symbolic unity in the new Warsaw, engaged in a rigorous publicity campaign using photographic images and ideological pronouncements cloaked in nationalist rhetoric to reshape popular perceptions of the city and remould the collective consciousness. In addition, socialist realist architecture adhered to strict stylistic protocols for new construction that drew on prototypes from Warsaw's historic landscape, which themselves underwent some form of revision in the course of their reconstruction.

It is no wonder then that the reconstruction scheme of 1949, co-authored by many of Poland's leading architects who had been active in the inter-war years, reflected modernism's implicit faith that rational organizational strategies operating at the regional scale combined with functionalist utilitarian solutions at the local level were capable of eliciting real social change; this despite the fact that ideological imperatives reflected in the rhetoric of socialist realism precluded such an overt acknowledgement of the reconstruction plan's modernist roots.[38] Thus the true coup of socialist realism in the Polish context was not merely its positioning as the superior and definitive substitute for modernism, but its successful conflation of the post-war ideological rejection of so-called functionalist cosmopolitanism and bourgeois capitalist architecture with popular perceptions of the pre-war urban landscape. More specifically, by convincing the Polish audience largely through the photographic medium of the privileged status accorded to the nation in the symbolic vocabulary deployed in the newly constructed Warsaw as well as in the city's resurrected historic districts, few could reject the legitimacy of the socialist regime and its architectural practices. This message was disseminated through the vehicle of photography in both exhibitions and the popular press along with a propaganda campaign that deployed a carefully crafted barrage of slogans and ideologically appropriate evocations of the nation. Thus, the socialist vision of the modern metropolis that promised to restore Warsaw's national cultural integrity necessarily entailed eliminating all material vestiges of the capitalist era, and, more significantly for Polish architects, it demanded expunging the perceived ideological and methodological patina of pre-war international modernism from the post-war Polish architectural discourse.

Notes

1. See Diefendorf, J.M. (ed.) (1990) *Rebuilding Europe's Bombed Cities,* Basingstoke: Macmillan; Diefendorf, J.M. (1993) *In the Wake of War: The Reconstruction of German Cities after World War II,* New York, Oxford: Oxford University Press; Appleyard, D. (ed.) (1979) *The Conservation of European Cities,* Cambridge, MA: MIT Press; Grebler, L. (1956) 'Europe's Reborn Cities', *Urban Land Institute Technical Bulletin* 28.
2. Kersten, K. (1991) *The Establishment of Communist Rule in Poland, 1943–1948,* trans. J. Micgiel and M.H. Bernhard, Berkeley, Los Angeles: University of California Press, pp. 441–67; Davies, N. (1984) *God's Playground: A History of Poland,* Vol. 2, New York: Columbia University Press, pp. 556–81.
3. For a summary of planning activities in Warsaw both during and after the war see Jankowski, S. (1990) 'Warsaw: Destruction, Secret Town Planning, 1939–44, and Postwar Reconstruction' in Diefendorf, J.M. (ed.) *Rebuilding Europe's Bombed Cities,* Basingstoke: Macmillan, pp. 77–93.
4. Bierut, B. (1951) *The Six-Year Plan for the Reconstruction of Warsaw,* Warsaw: Książka i Wiedza.
5. Ministry of Culture and Art (1945) *Warsaw Accuses,* Warsaw: Drukarnia Narodowa.
6. Kracauer, S. (1995) 'Photography' in *The Mass Ornament,* trans. T.Y. Levin, Cambridge, MA: Harvard University Press, pp. 47–63.
7. Ibid., p. 59.
8. 'Piękno Warszawy której już niema, a którą wskrzesimy', *Stolica* 1,1 (3 Nov. 1946): 6–7.
9. Ibid.
10. Ibid.
11. Ibid.
12. Ibid.
13. This key trope of Polish national identity has its roots in the epic poem *Pan Tadeusz,* written by Adam Mickiewicz in 1834 in the aftermath of the 1830 Uprising and following the period of partitioning (1772–1795) and the 1815 Congress of Vienna, when the territories of the Commonwealth of Poland-Lithuania were divided between Russia, Prussia and Habsburg Austria. See Mickiewicz, A. (1992) *Pan Tadeusz,* trans. K.R. Mackenzie, New York: Hippocrene Books.
14. Clark, K. (1999) *The Soviet Novel: History as Ritual,* Chicago: University of Chicago Press, p. 59.
15. Crowley, D. (2002) 'Socialist Spaces' in Crowley, D. and Reid S.E. (eds) *Socialist Spaces: Sites of Everyday Life in the Eastern Bloc,* Oxford: Berg, p. 8.
16. Gross, D. (1992) *The Past in Ruins: Tradition and the Critique of Modernity,* Amherst: University of Massachusetts Press, p. 58.
17. Durko, J. (1957) 'Le Musée D'Histoire de Varsovie', *Museum* 10,4: 258–67.
18. Ministry of Culture and Art, *Warsaw Accuses,* p. 3.
19. Syrkus, H. (1946) 'Warszawa oskarża -Warszawa żyje' in Barucki, T. (ed.) *Fragmenty stuletniej historii,* Warsaw: SARP, pp. 277–8.
20. Gropius, W. and Syrkus, S. (1946) 'Introduction and Keynote Speech', American Institute of Architects, Washington, Chapter convention, Library of Congress, Washington, DC, 16 Apr. 1946.

21. Quoted in Syrkus, H. 'Warszawa oskarża – Warszawa żyje', p. 278.
22. For an overview of the main events and personalities at the Athens Congress see Mumford, E. (2000) *The CIAM Discourse on Urbanism, 1928–1960,* Cambridge, MA: MIT Press, pp. 73–91.
23. Additional evidence of the attempt in the early post-war years to maintain continuity in the international modern architectural discourse comes from the first post-war CIAM conference, convened at Bridgwater, Somerset, in September 1947, where Helena Syrkus, the wife of Szymon Syrkus, was elected as a vice-president of the organization along with Gropius and Le Corbusier. In less than two years, however, this optimism amongst architects dramatically waned as the political-ideological East-West divide became increasingly unbridgeable. For more on the breakdown of relations between architects in the East and the West see Giedion, S. (1951) 'Post-War Activity of CIAM' in Giedion, S. (ed.) *A Decade of New Architecture,* Zurich: Editions Girsberger; and Giedion, S. (1958) 'Architects and Politics: An East-West Discussion' in *Architecture You and Me: The Diary of a Development,* Cambridge, MA: Harvard University Press, pp. 79–90.
24. Archiv Institute für Geschichte und Theorie der Architektur/Eidgenössische Technische Hochschule Zürich: Syrkus, Letter from Helena Syrkus to Dr. Martin Steinmann, 12 Mar. 1977.
25. Mumford, L. (1946) 'The New Warsaw', *New York Times,* 28 Apr. 1946: E10.
26. See Strzelecki, E. (1946) 'The Reconstruction of Warsaw', *Poland of Today* 1,5 (July): 4–7.
27. Ibid., p. 7.
28. Kotarbiński, A. (1945) 'Odbudowa czy Przebudowa', *Odrodzenie,* 56–57: 412–20.
29. Ibid., p. 413.
30. For a full version of the Athens Charter and a record of the proceedings of the CIAM IV Congress see 'Le IVe Congres International D'Architecture Moderne à Athenes "La Ville Fonctionnelle"', *Annales Techniques,* 15 Oct.–15 Nov. 1933.
31. Syrkus, S. (1947) 'Warszawa Przyszłości', *życie Osiedli Warszawskiej Spółdzielni Mieszkaniowej,* 2, p. 385.
32. Ibid.
33. Wołoszynowski, J. (1948) 'Nowa Warszawa', *Dziennik Literacki,* 38, p. 171. See also Kotańska, A. (1999) 'Dokumentacja fotograficzna wystaw: Warszawa, wczoraj, dziś, jutro (1938 r.), i Warszawa oskarża (1945 r.) w zbiorach Muzeum Historycznego m. st. Warszawy', *Almanach Muzealny,* 2: 291–313.
34. Kersten, *Communist Rule,* pp. 441–67; Davies, *God's Playground,* Vol. 2, pp. 556–81.
35. The People's Democracy was the official name of the Polish state until it was changed to the Polish People's Republic with the passing of the Constitution of 22 July 1952.
36. Minorski, J. (1949) 'Narada partyna architektów', *Nowe Drogi,* 3,3 (July–August): 139–42.
37. Bierut, *The Six-Year Plan,* pp. 39–40.
38. See Sigalin, J. (1986) *Warszawa 1944–1980: Z archiwum architekta,* Warsaw: Państwowy Instytut Wydawniczy. Józef Sigalin (1909–1983) was the director of BOS and chief architect of Warsaw from 1945 to 1950.

PART IV

Townscapes in opposition

11

CHARTING THE CHANGING APPROACHES TO RECONSTRUCTION IN FRANCE

Urbanisme 1941–56

Nicholas Bullock

Introduction

In 1948 the members of CIAM (Congrès Internationaux d'Architecture Moderne) gathered in Bergamo for the second post-war congress (CIAM VII) with a sense of anticipation about the opportunities that rebuilding Europe's war-torn cities would provide.[1] Journals like *Casabella* in Italy or the *Architectural Review* in Britain, longstanding champions of Modernism, were talking up the potential of Modern architecture's contribution. In Germany the Werkbund was already assembling an exhibition – its first since the war – to publicize the role of the New Architecture in the reconstruction of a new democratic Germany. Across Europe there was an expectation, even a general presumption, that some form of Modern architecture, rather than a return to the traditions of the past, would prove the natural order for post-war reconstruction.

France was no exception. Thumbing through the pages of progressive architectural journals like *Architecture d'Aujourd'hui* and *Techniques et Architecture* from the mid 1950s onwards, it would be easy to form the impression that the reconstruction of France's war-damaged towns and cities was being carried out in terms that broadly followed the principles adopted by CIAM at their Bergamo meeting (Figure 11.1).[2] This 'rational' approach to planning was based on the principles set out in Le Corbusier's *Charter of Athens* (1943): the separate zoning of the basic functions of the city – working, living, travel and leisure; the separation of the road network from the layout of buildings; and the use of freestanding buildings.[3] In place of the 'closed' form of so many cities like Paris or Berlin where tenements lined the streets and the majority of flats looked onto narrow light-wells or courts behind, CIAM's approach promised a radically different, 'open' form of development, exemplified by Le Corbusier's plan for Saint-Dié, with its vertical neighbourhoods and open city 'core'. From the late 1940s onwards, these ideas were actively promoted by the Ministry of Reconstruction and Urbanism (MRU) under Eugène Claudius-Petit, a committed supporter of Modern architecture.[4]

By the early 1950s the first fruits of this approach were already being published: Lurçat's plans for the rebuilding of Maubeuge, Roux and Pinguisson's proposals for the reconstruction of the Saar (Sarre in French) and Marcel Lods' flats at Sotteville-lès-Rouen, to name but a few.[5]

But this view of reconstruction, reinforced by the emphasis of recent studies of Modernism in France, distorts our understanding of the way in which French towns and cities were actually rebuilt after WWII, consistently underplaying the importance of an older tradition of French urbanism and planning.[6]

(A)

FIGURE 11.1 A widely publicized example of reconstruction in accordance with the principles of the Charter of Athens, Sotteville-lès-Rouen by Marcel Lods, 1948–55: (A) view of the slab-blocks; (B) and site plan

(B)

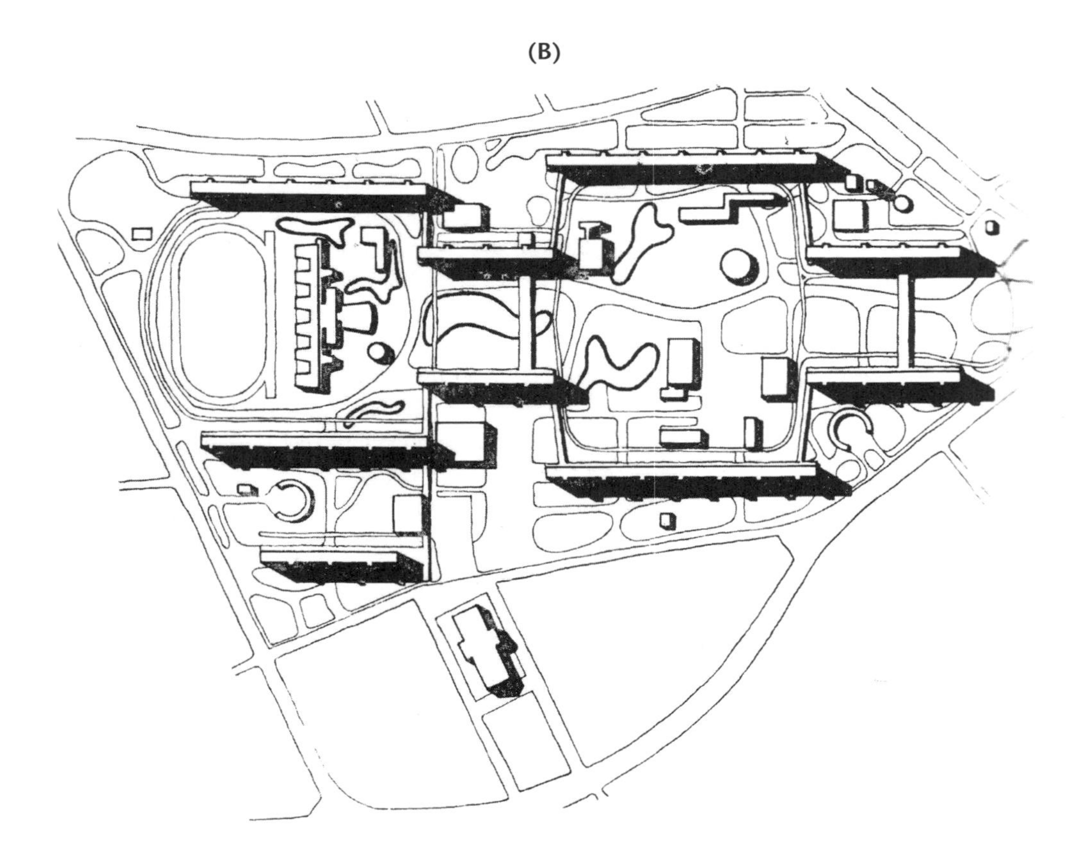

This tradition had its roots in the ideas debated before WWI by bodies such as the Musée Sociale and the Société Française des Urbanistes, in the designs for housing prompted by events such as the Fondation Rothschild's competition of 1905 and the discussions of the redevelopment of the Parisian fortifications.[7] During the inter-war years both the practice and the theory of French urbanism had developed at a gathering pace. The expansion of practice arose through the opportunities for French planners: their prestige, backed by the international reputation of the Beaux Arts, won them a number of major commissions abroad;[8] at home, more slowly, they were called upon to prepare the plans for the beautification and extension of towns and cities required by the Cornudet Act of 1919. The advance of theory was promoted by the establishment of the first courses to teach *urbanisme* and by the establishment in 1919 of what was soon to become l'Institut d'Urbanisme de l'Université de Paris. By the mid 1920s these ideas formed the mainstream of French thinking on urbanism. It was to this tradition, a set of overlapping approaches rather than any formal aesthetic, that urbanists and architects turned as they started in 1941 to prepare plans for the reconstruction of towns damaged in the German invasion of 1940. It would also provide the starting point for so much of what was actually built after 1945.

From its launch in early 1932, the evolution of this mainstream tradition in all its variety can be followed in the pages of the journal *Urbanisme*.[9] It was to become – and indeed still remains – the journal of record of French urbanism, of the tradition of town design and of larger-scale town and country planning, providing a continuous record of both practice and the debates on urbanism.[10] By following the way that the journal responded first to the war, then to the invasion and, finally, to the new regime, we can understand the way that planners adapted the older tradition in response to the wartime challenges of preparing for reconstruction under the German occupation. With the Liberation and the start of post-war rebuilding, we can trace the way in which the older tradition came to be supplanted by the new CIAM orthodoxy. In the mid 1950s, as the editors looked back over ten years of reconstruction, we can judge quite how much had been achieved by those working within this tradition before its eclipse.

Urbanisme, the pre-war years

From the start, *Urbanisme* brought together in a single publication the principal strands of the French debate on urbanism. In the introduction to the first issue, the two editors, Henri Prost (then preparing the first plan for the Paris region) and Jean Royer (Prost's former student and his assistant on the Paris plan), presented the agenda of the new journal as being shaped by a number of different institutions representing the different 'voices' in the contemporary debate.[11] Poëte's Institut d'Histoire, de Géographie et d'Economie Urbaines de la Ville de Paris and the Musée Sociale, then directed by Georges Risler, were given pride of place as 'patrons' of the journal, and through them the editors could count on the active support of key figures associated with these institutions: Auguste Bruggeman, Louis Bonier, Louis Dausset, Henri Sellier and others who had done so much for urbanism and the wider cause of 'social hygiene' in France since 1900.

A measure of the width of the journal's interests is the editors' inclusion of a list of editorial board members that included individual 'notables' and representatives of five other institutions. First of these was the Union des Villes et Communes de France, whose presence signalled the determination of the editors both to engage with the practice of urbanism and to make a case to those involved in the administration of local authorities for the practical value of planning.

Equally important in terms of day-to-day applications was the inclusion of the Association Française pour l'Amelioration de l'Habitation, which ensured that the journal retained active links with the Habitations à Bon Marché associations, the *cités-jardins* and the social housing movement. Representation of the various areas of technical expertise linked to urbanism was provided by the Société Française des Urbanistes, whose members included not just the Grand Prix de Rome-winning architects like Hébrard, Gréber and Jauselly – and Prost himself – but engineers like Forestier and landscape designers like Redont.[12] To ensure a balance between the grand old men of the older generation of urbanists, active before the war, and the younger generation, trained post-war, the editors included a representative of the Société des Diplomés de l'Institut de l'Urbanisme de l'Université de Paris. The final institution represented on the editorial board was the Institut International de l'Urbanisme Colonial, a reminder of the importance for French urbanists of the opportunities provided by the need to plan the principal cities of France's empire and of the international prominence given to French urbanism by the work abroad of Agache, Lyautey, Hébrard and others.

From 1932 until the war, Royer, working as executive editor with Prost, chair of the editorial board, produced a flow of articles that not only broadly matched this impressive spread of interests but also provided a balance between more pragmatic concerns such as the planning of the Paris region and more theoretical or academic articles that served to anchor the journal's core identity. Pre-war, *Urbanisme* stood for an approach to planning that drew on both science and the arts, and, as might be expected with so strong a debt to Poëte, on a keen sense of history and the particularities of place. The journal's interests in social reform, inherited from the Musée Sociale and Henri Sellier, were balanced by its publication of the 'imperial' designs of Grand Prix de Rome winners like Gréber. More than most French journals of the time, *Urbanisme* seemed open to developments abroad, to planning in the USA and Britain and, exceptionally, to what might be learnt from the German tradition of planning.

But though the editors protested their desire to be inclusive, not all voices were equally welcome. The development of Modernist ideas remained unreported.[13] The housing built at Drancy by Beaudoin and Lods, one of the very few Modernist projects published before the war, was included under the rubric of *cité-jardins* of the Seine, not Modern architecture. There was, for example, no constructive comment on or review of Le Corbusier's Ville Radieuse, nor any coverage of the various submissions that he made to competitions reviewed by the journal. By contrast with the *Architectural Review, Urbanisme* showed no pre-war interest in the New Architecture.

The new opportunities created by the war

The outbreak of war and the German invasion of 1940 transformed the world that *Urbanisme* sought to address, and with it the agenda of the journal. The journal ceased publication after the April–May issue of 1939 and did not reappear until the January–May 1941 issue. A central task of the journal during the 1930s had been to make the case for planning. Now, with the destruction of so many towns and cities from the Nord to the Loire, the priorities were quite different: how were they to be rebuilt? How was France to avoid missing the opportunities that had been let slip in 1918?

When the journal reappeared in 1941 one of the editors' first priorities was to position the journal positively vis-à-vis the new regime. In part this was achieved through the publication of a series of articles that sought to endorse the regime's call for a moral renaissance that would transform France. Typical of these were André Véra's extravagant exhortations in 'Manifesto for the Renewal of French Arts', pledging the support of the arts, from interior design to landscape gardening, from sculpture to architecture, to the 'revolution nationale' and the revitalization of French culture.[14] But more important – and central to its own aims – was the journal's support for the legislation passed during 1940 and 1941 that created the machinery for a more radical approach to reconstruction than had been possible in 1918.[15]

Central to Royer and Prost's support for this legislation was the argument that the powers now made available for rebuilding by the new État Français were those that urbanists and planners had demanded for so long. At long last, the journal could welcome proposals that placed the needs of the collective ahead of the rights of the individual. Moreover, for its part, the regime recognized the contribution that the journal had played in preparing the ground for this legislation. Lehideux, writing in the November 1941 issue of *Urbanisme* of the government's ambitions for a 'renaissance' of France that would rekindle the moral and physical health of the nation, was happy to salute *Urbanisme* as a leading champion of the state's new powers.[16]

Eager to keep the momentum moving, the journal devoted its pages to documenting both the new legal machinery of reconstruction and what was being achieved in practice. The November 1941 issue of the journal, the second after its reappearance, contained the New Charter for Urbanism, presenting the government's intentions and a long comparison by Royer, under the title 'Reconstruction 1941', of the shortcomings of reconstruction in 1918 with the possibilities now available.[17] The December issue was devoted entirely to setting out, along with an extensive commentary, the details of the new powers of *remembrement,* enabling the authorities to redistribute land holdings in the interest of an orderly and rational reconstruction.[18] The March 1942 issue was given over to a discussion of additional planning powers, the procedures for replanning, rerouting traffic, opening up public spaces and making other improvements as part of the programme of reconstruction.[19]

Alternating with these issues on the new planning powers were others devoted to illustrating what reconstruction might achieve in practice. Over the next two years, issues on the machinery of reconstruction were followed by those devoted to reconstruction in the different Départements that had suffered heavily during the invasion of 1940. The journal opened its presentation of practice with the first reconstruction plans to be announced, those for the Department of the Loiret.[20] With Royer appointed *urbaniste en chef* for the Department, plans for the rebuilding of the towns along the Loire, the smaller towns like Gien, Chateauneuf and Sully, and Orléans, the largest town in the area, exemplified a blend of past and present that was to characterize so many wartime plans and seemed so well suited to the values of the new regime (Figure 11.2).

The planners' deference to the past was evident in the way that the proposals for each town were framed with a brief historical summary of its development, and the plans were for the most part quite modest, for remodelling and improvement rather than radical reconstruction. Thus, at Sully, Royer's ambition was, first, as far as possible, to help the traffic along the bank of the Loire and from Montargis in the north to Bourges in the south to by-pass the town; and, second, to open up the square in front of the church – now freed of traffic – to provide more effectively for the public life of the town. In Orléans, too, replanning was limited to easing the congestion by opening up through routes outside the immediately central area, to rebuilding slum property and to using the new powers to redistribute land ownership in order to create a series of open spaces to display more effectively the city's surviving historic monuments.

(A)

FIGURE 11.2 One of the first plans for reconstruction after the invasion of 1940: (A) Laborie's plans for the rebuilding of Gien; (B) the town as rebuilt; (C) changes to the traffic system

(B)

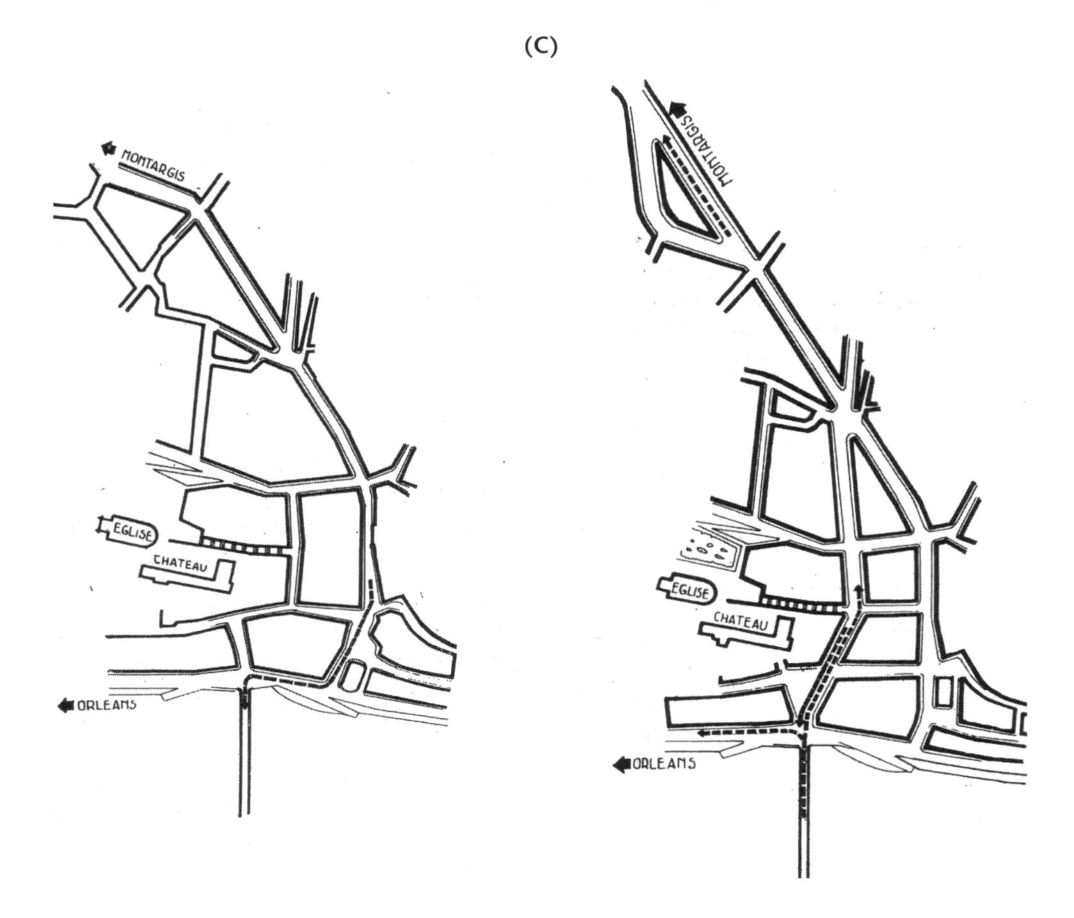

Approved during the summer of 1941, the plans prepared by Royer and Kérisel for these reconstructions in the Loiret were to be of exemplary importance for wartime plans for rebuilding other towns in France. Their influence did not result only from the publicity that followed their publication in *Urbanisme* but was further enhanced by the central positions that the two men came to occupy in the administration of reconstruction: Kérisel as technical director of the Commissariat à Reconstruction Immobilière and Royer as head of the under-secretariat responsible for reviewing all plans for reconstruction submitted from across France.

Important, too, were the links established in these early proposals in the Loiret between the planning and the architecture of reconstruction. From the start, as was clear from the deliberations of the Comité National de la Reconstruction, there was little to distinguish the role of the urbanist from that of the architect.[21] Urbanism was very much a visual affair, design at the scale of the town rather than that of the individual building. The strength of these links was illustrated from the very first issue of *Architecture Française,* a journal that expressly claimed to champion in architecture the values of the regime. Like Royer's proposals for improving and repairing the fabric of the towns, Laprade's drawings for Gien exemplify the same concern with history, a comparable conservatism of approach, the same emphasis on regionalism. This is evident in the drawings of vernacular buildings that he presents as a preface to the results of the competition for the reconstruction of the town's riverside shops and houses (Figure 11.3).[22]

(A)

FIGURE 11.3 The architectural vocabulary favoured by Vichy: (A) traditional building in Gien; (B) one of the winning schemes in the competition for rebuilding shops in the Loiret, 1941, by Laprade, Bazin, Neau and Delval

(B)

The winning designs may reflect the regime's conservative emphasis on tradition and the crafts, but the forms were offered as more than mere pastiche. Like the plans for reconstruction, the architecture was to be shaped by local materials, local traditions of construction and a realistic approach to rebuilding in straightened circumstances. As early as February 1941, André Leconte, the regime's *architecte en chef,* codified the architectural and planning experience drawn from these first reconstruction initiatives as the Charter of the Architect-Reconstructor in order to provide guidance for reconstruction elsewhere in France.[23]

Over the next two years, *Urbanisme*'s publication of plans for reconstruction, presented with this strong regional emphasis, was extended to cover departments like the Aisne, the Oise, the Somme and others that had suffered in the fighting of summer 1940.[24] The approach favoured by wartime urbanists looked very different from the Beaux Arts schemes which had figured in the journal in the early 1930s. In place of imperial compositions on tabula rasa sites, the new approach was low-key and respectful of the existing grain of what were for the most part the late medieval or early modern 'cores' of small market towns. Plans of the time suggest a Sittesque concern with the shaping – or 'repairing' – of public spaces. They were informally rather than symmetrically composed to create contrasting views that drew the eye in the Picturesque manner down narrow streets ending with church towers or other points of visual emphasis. The impression of the emergence of a new urban vocabulary for wartime plans is further heightened by the accompanying architectural proposals, which were generally deferential to the form of neighbouring buildings and local traditions of construction.

Typical of the wartime plans published in *Urbanisme* are those drawn up by Gaston Bardet, Marcel Poëte's son-in-law. He had started to teach at l'Institut d'urbanisme de l'Université de Paris before the war and was to become both a regular contributor and, in 1941, a member of the editorial committee of the journal. His plans for Vernon, for example, are rooted in an understanding of the historical development of the town.[25] Here, by rebuilding war-damaged blocks in the centre of the town and slightly changing the alignment of the roads, Bardet was able to solve the intense problems of traffic congestion created at the bridge over the Seine in a way that enhanced the qualities of the existing fabric.

Equally important, Bardet's articles reveal something of the journal's wider social concerns during the war years.[26] For Bardet, whose conservative Catholic Humanism might share some of the social priorities of the Vichy regime, the formal emphasis of much pre-war urbanism, whether the elaborate compositions of the Beaux Arts or the reductive *machinisme* of the Modernists, was unacceptable. In contrast to the 'nihilistic' and lifeless formalism of these approaches, Bardet had been schooled by Poëte to think of the city as a living organism whose form was a reflection over time of the changing social, economic and political life of a community. In preparing his plans for reconstruction, Bardet surveyed, analysed and then visually represented the activities of the towns on which he worked in order to reveal what he called their 'social topography'. For Bardet, the task of the urbanist was to respond to the pulse of the city's activities, to propose plans that answered as directly as possible to the flux of urban life, to the patterns of pedestrian movement or the flows of traffic. In doing so, the planner was also to encourage the formation of those qualities of community that were so necessary for the individual to flourish, and were the key to a humane urbanism that would hasten the physical and spiritual regeneration of society.

Liberation and new priorities for *Urbanisme*

In June 1944, the programme of reconstruction, laboriously put in place after the invasion of 1940, was brought to a halt by the Allied landings. Reconstruction had now to wait upon the end of the war. But the period following the Liberation was not lost. Preparations for reconstruction continued, and in November 1944 Raoul Dautry, a member of *Urbanisme*'s editorial board from the start, was appointed Minister of Reconstruction and Urbanism. As in so many other areas of French life, the Liberation signalled both a new beginning and the continuation of developments under Vichy.[27] Despite the creation of a new ministry to deal with all aspects of reconstruction, the new minister maintained largely unchanged the administrative machinery that had been created in 1940–41 and most of its senior staff.

Urbanisme, too, revealed something of the same mixture of change and continuity. The last wartime number of *Urbanisme* appeared in May 1944, and the journal did not reappear until the summer of 1945. When it did, it was clear that the journal was already beginning to change. Continuity was evident in the editorial line-up: Royer and Prost remained at the helm, and a significant number of members of the editorial board survived, including a number of those who had joined before the war.[28] Véra's vaporous essays continued to be published, and the old format of reviewing plans for reconstruction by Départements continued, with plans that still looked broadly similar to those prepared in the earlier years of the war.[29] Change, however, was evident in the departure from the editorial board of André Muffang – too closely identified with Vichy – and, more important, in the impact of the newly established MRU:[30] Raoul Dautry, the new minister, became the journal's honorary president, and key officials at MRU like André Prothin made their first appearance as members of the editorial board. However, the new minister had pragmatic views on reconstruction and catholic views on architecture so that the new ministry was, at least initially, flexible in its choice of urbanists and architects. It consulted locally and appointed both traditionalists and Modernists.

The issues of *Urbanisme* published immediately after the war, on a quarterly rather than a monthly basis,[31] suggest a loss of purpose and the beginnings of a separation between the choice of material published and the rapidly increasing number of new plans for the towns and villages that had suffered in the last year of the war. Most obvious, despite Royer's retrospective claim to have been open to every point of view, was the editors' evident unwillingness to publish plans that broke with the journal's wartime approach.[32]

However, the number of Modernist plans was growing. This was due in part to the greater freedom of layout, especially the ability to separate the building block from the street line provided by Vichy's legislation on *remembrement.* Architects and urbanists could now propose designs in the 'open manner'. Plans as architecturally diverse as Auguste Perret's plan for Le Havre, André Lurçat's plan for Maubeuge and, at a smaller scale, Stoskopf's proposals for Ammerschwihr illustrated the new freedom.[33] This same legislation also provided Le Corbusier with the means to realize the vision he had set out in *La Ville Radieuse.* When first published in 1943 his Charter of Athens had had little effect, but the post-war popularization of these ideas in *A Propos d'Urbanisme* and in *Manière de penser l'urbanisme,* backed by his unsuccessful project for the rebuilding of Saint-Dié, proved much more influential.[34] They came at a time when the freedom offered by *remembrement* was more widely recognized and the scale of reconstruction in many towns much larger. From 1945 onwards the new 'open manner', often linked with Modern architectural forms, was increasingly visible: MRU chose to place Lurçat's radical plans for rebuilding Maubeuge centre stage in the Exhibition of Reconstruction and Urbanism that opened at the Grand Palais in the spring of 1946.[35] But to these important developments *Urbanisme* appeared indifferent throughout 1946, a year of crucial importance for the planning of reconstruction.

Then, in the spring of 1947, the journal changed direction, replacing at last any suggestion of continuity with the values of the Vichy years with an optimistic embrace of Modernism. In the March issue Prost's name disappeared from the masthead. In the April issue not only was Royer's name missing,[36] but the editorial entitled 'Resurrection', written by the new editor, Léon Moine, declared the death of the journal's old approach and promised in its place a new orientation and renewed engagement with the developments of the moment.[37]

The journal's new orientation was spelt out in the May issue. The introduction by Claudius-Petit, the extended description of Roux proposals for urbanism in the Saar and the call by Marcel Lods for the widespread application of the principles of the Charter of Athens represented a fanfare of support for Modernism. Though Claudius-Petit was not to become minister at MRU until the following September he was already, as deputy for the Loiret and a key member of the Assemblée Nationale's Commission de la Reconstruction et de l'Urbanisme, an influential figure on issues relating to reconstruction. He was known as a strong advocate of Modern architecture and a loyal (and long-suffering) supporter of Le Corbusier and the embattled construction of the Unité d'Habitation in Marseille, and his introductory article, 'A Contribution to Urbanism in France', was a plea for a bold approach governed by a few simple rules – by implication the Charter of Athens.[38] In opposition to those who looked to the past, he urged a firm belief in the promise of a new architecture as firmly rooted in the present as the great styles of the past had been in the values of their contemporary society: 'We live in an extraordinary period. The tired and timid amongst us may judge it wanting. But, despite its errors and its shadow moments, we must believe in it'.[39] Modern architecture, he argued, was not just an affirmation of faith in the promise of the future; it was also a continuation of the great French tradition adapted to the technologies and demands of post-war France.

The publication of Roux and Pinguisson's plan for the reconstruction of the Saar appeared to answer Claudius-Petit's encouragement to planners to be bold and forward-looking (Figure 11.4).[40] The regeneration the Saar was far from being a local matter. Crucial to the

(A)

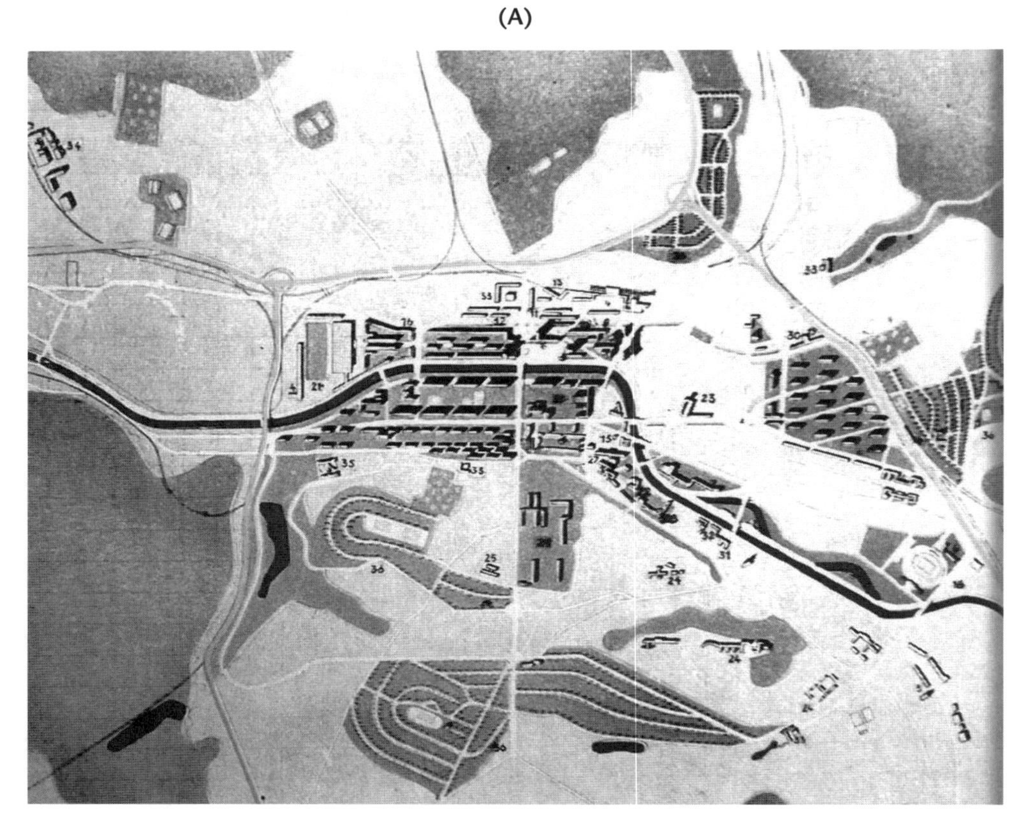

FIGURE 11.4 Reconstruction according to the principles of the Athens Charter, Roux and Pinguisson's proposals for the Saar, 1949: (A) a general plan showing the zoning of the town; (B) a sketch of the proposals for the riverside area

(B)

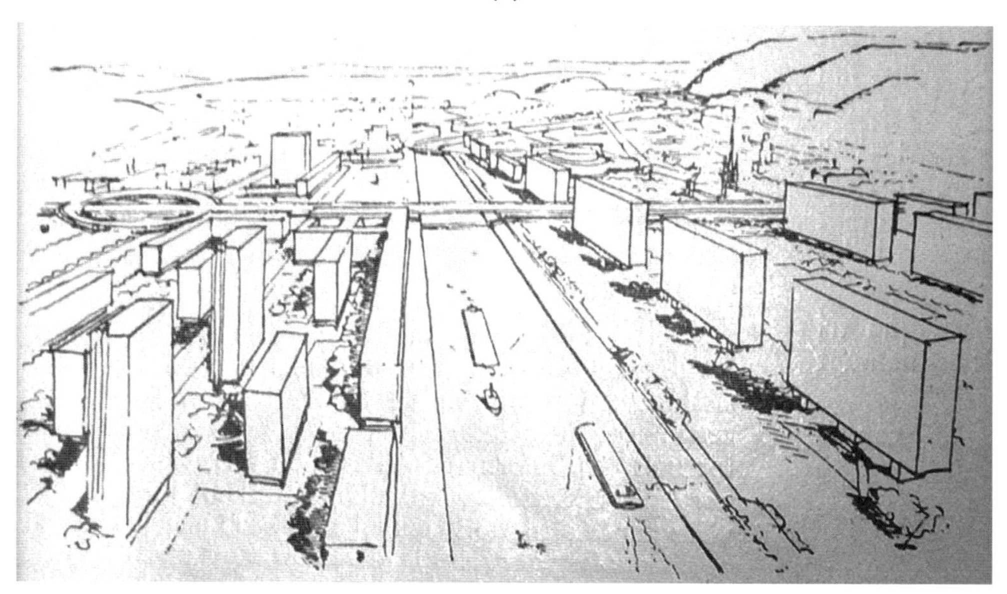

expansion of coal and steel production on which France's first National Plan depended, the modernization of the Saar was also intended as a demonstration of the benefits that French culture could offer a region which was yet to choose whether it should lie within Germany or France.[41] The form of the plan, and the way in which it was presented, emphasized the influence of the Athens Charter: the creation of different zones for each land use, the separation of these from the road networks for different kinds of traffic and the generous provision of a series of cultural, sporting and leisure facilities. With its large-scale housing blocks and clearly identifiable public buildings, the proposals exemplified radical Modern urbanism and the new 'open order' of development now favoured by MRU.

The journal's newfound identification with Modernism was further underlined by Marcel Lods' endorsement of the value of the Athens Charter as an invaluable aid to the hard-pressed urbanist in dealing with reconstruction committees unaware of what reconstruction and the new approach to 'open planning' might make possible.[42] Lods, whose Modernist credentials had been firmly established in the 1930s, presented the document as 'purely objective', drawn up without stylistic preconception and self-evidently appropriate to the needs of the moment: 'The greatest compliment that can be paid to this document is that, for so many questions, it offers extremely simple answers which lead the reader to conclude, "Yes, this is clearly the way to do it", and the problem finally solved'.[43]

Gaston Bardet's infuriated response to Lods' disingenuous presentation of the Athens Charter is a reminder of the strength of the disagreements between the supporters of the old and the new directions of the journal.[44] Accusing Lods of grossly simplifying the nature of urbanism, Bardet, styling himself 'David' in opposition to the Goliath of Modernism, denounced the Charter in forthright terms: 'The Athens Charter is completely out of date. It fails to take account of the social nature of the town now recognized by urbanists everywhere. It provides false and actively misleading evidence on high-rise housing, which is for CIAM the touch-stone of modernity but which, as Frank Lloyd Wright would say, is in reality the gravestone of civilisation.' Le Corbusier he denounces as the 'high-priest of the machine', a self-promoting charlatan, 'a man struggling beyond the scale of his competence who seeks to dazzle us with designs for sky-scrapers while born to do little more than decorate the watch cases of the middle classes'.[45]

The disagreements that resulted from the direction set for the journal in March 1947 flickered on. Bardet, a regular contributor during the war, ceased to write for the journal and left the editorial board.[46] But others were prepared to accept the new orientation. Royer returned to the magazine later in the year as *rédacteur en chef,* but the journal appeared to be limping along with little sense of direction. Only one issue was published in 1948 before the journal again ceased publication, with editorial policy mired in the dispute.

Conclusion: another new beginning

In early 1950, after a break of almost two years, the journal reappeared. The divisions of two years before appeared to have been resolved: Prost, at 76 the grand old man of French planning, was again described as the president of the Editorial Committee, while Royer seems to have taken on the more active editorial role of director. But if Prost was there on honorary terms and Royer was now actively in charge, the balance of articles reflected a shift towards Modernism and the priorities of MRU. Claudius-Petit, now minister and keen to extend MRU's support for Modernist planning and architecture, introduced the first of the new series and the new priorities. The article,'Programmes and the plastic', by Pierre Dalloz, Claudius-Petit's right-hand man at MRU, encouraged an approach to reconstruction along the lines urged by CIAM.[47]

From the start of the new series, the bias of the journal favoured the new approach. During 1951, articles on the reconstruction of Marseilles not only covered the rebuilding of the port and Pouillon and Perret's work in the centre, but illustrated Le Corbusier's Unité d'Habitation still under construction.[48] The presentation of reconstruction in Lille was prefaced by illustrations of the radical design for the Foire Internationale with its shining metal external skeleton by Paul Herbé, Maurice-Louis Gauthier and Jean Prouvé.[49] The year ended with the publication of Dubuisson's project for SHAPE village at Saint-Germain, one of the projects that made the most successful architectural use of the new Camus system of industrialized building being actively promoted by MRU.[50] Indeed the journal appeared keen to publicize the activities of MRU, covering the winning scheme for the ministry's 1951 competition for the construction of 800 flats for the Cité Rotterdam in Strasbourg, an important success in making the case for the industrialization of housing construction.[51]

With the dust now settled on the editorial scuffles over policy of the late 1940s, *Urbanisme* could again lay claim to be the unbiased journal of record for urbanism, architecture and planning in France. Writing in 1952 and looking back over 20 years of almost continuous publication, Royer could emphasize the continuities, airbrushing out any record of dissent and division, and claim that the journal's editors had always taken an inclusive view of French urbanism:'Determined not to become the house journal of any "school" or of any individual, the journal remains what it has been for twenty years, a crossroads, a point of encounter for people from very different backgrounds with very different points of view but united in their aim of improving the collective life of their fellow men'.[52]

The journal's review in 1956 of the achievements of reconstruction after ten years offered a balanced selection of projects, providing some substance to Royer's claim to be open to all the different currents of urbanism.[53] The 66 projects, chosen from the 2,000 separate projects of reconstruction across the country, do indeed represent a wide variety of different approaches to reconstruction in France. They include projects traditional and Modern. The former are represented by projects of the early war years like Gien, with its concern to rebuild in a way that was sympathetic to the remaining historic fabric of the town, but also by projects of the post-war years like Arretche's reconstruction of the old town of Saint-Malo, combining industrialized building techniques with local materials – in this case granite – to recreate the scale and the silhouette of the eighteenth-century core. The Modernists' projects include Vivien's proposals for the radical transformation of the centre of Boulougne with the replacement of the fabric of the old port by the long slab-blocks marching along the bank of the estuary, and the even longer blocks of Lods' programmatic application of the Charter of Athens at Sotteville-lès-Rouen (Figure 11.5).

As the 1956 retrospect of reconstruction showed, Royer was keen to make a virtue of the journal's inclusive catholicism.[54] But the turbulent record of the publication during the late 1940s and early 1950s reveals more faithfully than Royer's editorial gloss the divisions of opinion and the hostility that marked the terms in which CIAM's vision of the 'rational city' came to supplant the older tradition of the 1930s and the wartime years.

But if in France, as elsewhere in Europe, the opportunities created by reconstruction had favoured Modern architecture and CIAM's approach to modernizing the city, this advantage

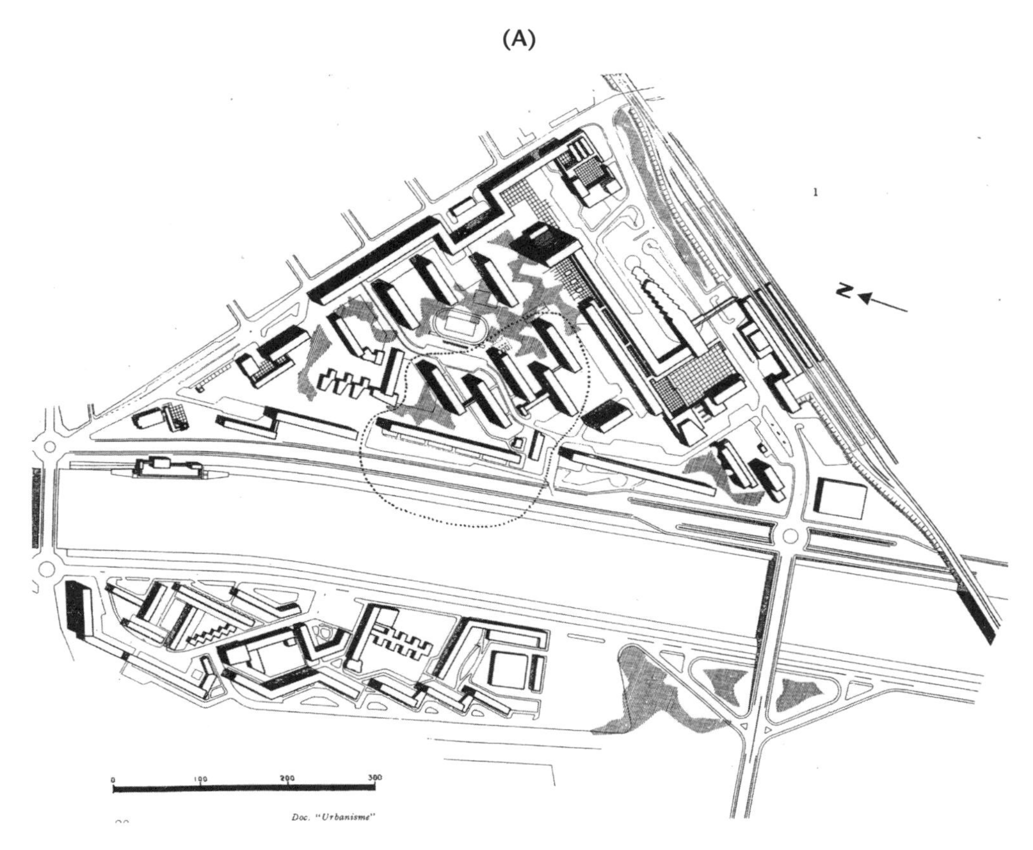

FIGURE 11.5 The Modernist architecture of reconstruction: Pierre Vivien's rebuilding of the port area of Boulogne, 1951–56: (A) site plan; (B) a view of the new housing from across the port

(B)

was soon to be challenged. As the plans for the new 'open' developments for cities like Angoulême, Boulogne and Rouen were taking shape, a younger generation was beginning to challenge CIAM's 'rational' approach to the city. At the CIAM meeting in Aix, the 'youngsters', Alison and Peter Smithson, and Aldo van Eyck, were already questioning the principles enshrined in the Athens Charter.

In shaping their urban ideal, the younger generation no longer looked for inspiration to the city of freestanding blocks set in open space, to Le Corbusier's Unité d'Habitation in Marseilles or to his plans for Saint-Dié. Instead, they turned to the qualities of enclosure and the making of place that they found in the typology of the traditional city. They welcomed, as did the editors of *Urbanisme* and those familiar with the older tradition of French *urbanisme*, the way in which the streets and squares of Europe's older towns and cities encouraged those qualities of association and community that were central to the everyday experience of urban life.[55] Though their architectural vocabularies were very different, the French urbanists of the 1930s and 1940s and the radicals of Team X shared a view of the essential qualities of town and city. In a number of key respects each had more similarities with the other than with the Modernists, whose vision of the rational city had come to dominate the debate on reconstruction in the decade after the war.

Notes

1. The role confidently expected for the New Architecture is conveyed in Sigfried Giedion's 1951 international survey, *A Decade of Contemporary Architecture,* Zurich: Editions Girsberger.
2. See for example the first major issue of *Architecture d'Aujourd'hui* dedicated to reconstruction, 'La Reconstruction en France', 8 (Sept–Oct 1946).
3. Le Corbusier (1943) *La Charte d'Athènes,* Paris: Pion.

4. Claudius-Petit's role at MRU is discussed in Pouvreau, B. (2004) *Une politique en architecture, Eugène Claudius-Petit 1907–1989,* Paris: Le Moniteur, especially Chapter 8.
5. See for example the numbers of *Architecture d'Aujourd'hui* devoted to the progress of reconstruction: 'Reconstruction France 1950', 32 (Oct–Nov 1950); and 'Contribution française à l'évolution de l'architecture', 46 and 47 (Feb–March and March–April 1953).
6. For example Lucan, J. (2001) *Architecture en France (1940–2000), histoire et théories,* Paris: Le Moniteur; and Abram, J. (1997) *L'Architecture Moderne en France, du chaos à la Croissance 1940–1966,* Paris: Picard.
7. For an introduction to its origins see Sutcliffe, A. (1981) *Towards the Planned City, Germany, Britain, the United States and France 1780–1914,* Oxford: Blackwell, pp. 134–62; for developments after 1940 see Newsome, W.B. (2009) *French Urban Planning,* New York: Peter Lang.
8. Jacques Gréber's design for the Benjamin Franklin Parkway in Philadelphia (1917) exemplifies the success of French urbanists abroad.
9. A number of journals provided coverage of urbanism and planning issues before the launch of *Urbanisme.* The principal architectural journals, such as *Architecture,* regularly gave space to matters of urban design, while other journals, for example *La Vie Urbaine* published by the l'Institut d'urbanisme de l'Université de Paris, addressed wider social and economic questions.
10. The role played by *Urbanisme* is closer in Britain to that of the *Architectural Review* than to the more technical and professional bias of the *Town Planning Review.*
11. Henri Prost (1874–1959), trained as an architect, was France's premier urbanist of the inter-war years; Jean Royer (1903–81), also trained as an architect, was editor of the École Spéciale's journal *Le Maître d'Oeuvre,* published 1926–30, which regularly featured questions of urbanism. Royer taught at the l'IUUP, where he was later to become a professor.
12. La Société Française des Urbanistes was founded in 1911 by D.A. Agache, M. Auburtin, A. Bérard, E. Hébrard, L. Jaussely, A. Parenty, H. Prost (architects), J.C.N. Forestier (engineer and landscape designer) and E. Redont (landscape designer).
13. Lacroix, for example, as director of the Union des Villes et Communes de France, was keen to draw a distinction between the realistic agenda of *Urbanisme* and mere Modernist speculation, a dismissal aimed no doubt at Le Corbusier's utopian planning projects such as his Voisin Plan for Paris, 'Considérations générales en matière d'introduction' (1932) *Urbanisme,* 1 (April), p. 3.
14. Véra, A. (1941) 'Manifeste pour le Renouveau de l'Art Français', *Urbanisme,* 72 (Oct–Nov): 53–59.
15. See particularly the Nov 1941 issue; Royer, J. (1941) 'Reconstruction 1941', *Urbanisme,* 72 (Oct–Nov): 64–100.
16. Lehideux, F. (1941) 'Reconstruction . . . Renaissance', *Urbanisme,* 72 (Oct–Nov): 50–51.
17. Moine, L. (1941) 'Vers une Nouvelle Charte de l'Urbanisme', *Urbanisme,* 72 (Oct–Nov): 60–63; see also note 15 above.
18. 'La Recherche du Parcellaire et sa Redistribution' (1941) *Urbanisme,* 73 (Dec): 102–143.
19. 'La Redaction des Programmes d'Aménagement' (1942) *Urbanisme,* 76 (March): 95–146.
20. 'Le Val de Loire' (1941) *Urbanisme,* 71 (Jan–May): 10–35.
21. See the transcripts of meetings of the Comité National de la Reconstruction reproduced in Kopp, A., Boucher, F. and Pauly, D. (1980) *France; L'Architecture de la Reconstruction, 1945–1955,* Paris: ARDU, pp. 54–58.
22. *Architecture Française,* 1 (1), p. 30.
23. Leconte, A. (1941) 'Charte de l'architecte-reconstructeur', *Architecture Française,* 2 (3) (Feb): 1–4.
24. The reconstruction of the Oise was documented in No. 77 (April 1942); the Seine-Inférieure in No. 80–81 (July–Aug 1942) and No. 82 (Oct 1942).
25. 'Vernon' (1943) *Urbanisme,* 88 (March): 58–61.
26. Bardet, G. (1943) 'Connaisance de la Ville', *Urbanisme,* 92–93 (July–Aug): 149–55; see also Bullock, N. (2010) 'Gaston Bardet, Postwar Champion of the Mainstream Tradition of French Urbanisme', *Planning Perspectives,* 25 (3): 347–63.
27. The transition is described by Voldman, D. (1997) *La Reconstruction des Villes Françaises de 1940 à 1954, Histoire d'une Politique,* Paris: Harmattan, Chapters 4–5.
28. Although some of the original members, for example Henri Sellier, had died, 22 out of the 45 board members remained.
29. See for example Vera, A. (1945) 'Opportunité de l'Urbanisme', *Urbanisme,* 105–06: 11–44; and the plans for the Moselle presented in *Urbanisme* (1946) (111–12): 44–54.
30. For Muffang's departure see Voldman, *La Reconstruction des Villes Françaises,* 124–25.

31. Possibly through lack of paper and ink, though other journals kept going, but more likely because of editorial disagreements.
32. Royer, J. (1952) 'Vingt ans après', *Urbanisme,* 21–22: 2–5.
33. For Stoskopf's plan for Ammerschwihr see (1949) *Techniques et Architecture,* 9 (3–4): 46–52.
34. See note 3.
35. 'L'Exposition de la Reconstruction' (1946), *Urbanisme,* 109: 132–33.
36. This may be explained in part by Royer's duties as *urbaniste en chef* for Orléans, where the arduous task of implementing the plans drawn up during the war was gathering pace.
37. Moine, L. (1947) 'Urbanisme d'Avril: Résurrection', *Urbanisme,* New Series, 2 (16) (April), p. 2.
38. Claudius-Petit, E. (1947) 'Une contribution à un urbanisme français', *Urbanisme,* 115 (New Series, 3) (May): 80–85.
39. Ibid., p. 85.
40. 'Urbanisme Français en Sarre' (1947) *Urbanisme,* New Series, 3 (May): 86–110.
41. Ibid., p. 102. For an account of the larger context of the replanning of the Saar see Baudoï, R. (1991) 'La reconstruction française en Sarre, 1945–1950', *Vingtième Siècle* (Jan–March): 57–65.
42. Lods, M. (1947) 'Urgence de la Charte d'Athènes', *Urbanisme,* New Series, 3 (May), p. 121.
43. Ibid.
44. Bardet, G. (1948) 'La réponse de David', *L'Architecture d'Aujourd'hui,* 16 (Dec): XIII–IX.
45. Ibid.
46. He stepped down from the editorial board in 1947 and, with the exception of a brief article in 1957, 'Caractère organique des tissues urbains' (*Urbanisme,* 54): 85–92, published no more articles in *Urbanisme.*
47. Dalloz, P. (1950) 'Programmes et plastique', *Urbanisme,* Vol. 19 (1–2): 16–29.
48. See the 1951 issue dedicated to Marseilles, *Urbanisme,* Vol. 20 (5–6), especially Hardy, A.P. 'L'habitat défectueux et la politique du logement': 29–34; and Malcor, R. 'La Réalisation du Plan d'urbanisme de Marseille': 48–52.
49. Catin, R. (1951) 'Le Quartier du Parc des Expositions', *Urbanisme,* Vol. 20 (9–10): 21–25.
50. 'SHAPE village' (1951) *Urbanisme,* Vol. 20 (11–12): 13–17.
51. 'Habiter autour d'un jardin; trois projets de E-E Beaudoin' (1951) *Urbanisme,* Vol. 20 (7–8): 5–8.
52. Royer, J., 'Vingt ans après', p. 5.
53. In 1956 the journal devoted a double number to the review of reconstruction, *Urbanisme,* 25 (45–48).
54. Royer, J. (1956) 'L'oeuvre de la reconstruction', in ibid., p. 181.
55. See for example 'Creations Urbaines', the special 1956 number devoted to the problems of the modern city, *Urbanisme,* 54.

12

BRUTAL ENEMIES?

Townscape and the 'hard' moderns

Barnabas Calder[1]

Introduction

> After the insufferable tedium of townscape, the dreary accumulation of public house chi-chi, and the insipid neo-Regency aesthetic with which we have been blanketed since the war, it is the most extreme relief to be allowed to recognise that English architecture is not necessarily degraded nor essentially corrupt.[2]

Thus wrote critic and theorist Colin Rowe to the *Architectural Association Journal* in 1957, to thank them for publishing an issue on the 1930s work of Connell, Ward and Lucas.[3]

The reference to 'public house chi-chi' was an unmistakable attack on the *Architectural Review:* it had published a special issue on the pub, with 57 pages devoted to a loving examination of the history of pub interiors. Furthermore, the journal's proprietor, Hubert de Cronin Hastings, had since the war been constructing from salvage a Victorian pub interior, 'The Bride of Denmark', in the basement of the magazine's offices at Queen Anne's Gate.[4] The other targets of Rowe's attack are equally clear. In collaboration with Thomas Sharp, the *Architectural Review* had adopted Townscape as a major campaign, advocating the application of Picturesque landscaping techniques as an aesthetic basis for town planning. Rowe's hostile hiss of 'insipid neo-Regency aesthetic' might call to mind the pinky-beige abstractions of terraces in the A. C. Webb illustrations of Sharp's Townscape city plans for Exeter and elsewhere.[5]

The leaders of the younger generation, by contrast, were then favouring increasingly stripped-down, hard-textured buildings; heavier use of concrete and industrial-looking brick; and expressive use of structure and construction techniques. Rowe, a leading theorist of these young hard modernists, was gunning for Townscape, Sharp and the *Architectural Review.*

Rowe was clearly one of those who, in Reyner Banham's words:

> [having] interrupted their architectural training in order to fight a war to make the world safe for the Modern Movement, tended to resume their studies after demobilization with sentiments of betrayal and abandonment. Two of the leading oracles of Modern Architecture [J. M. Richards and Nikolaus Pevsner] appeared to have thrown principle to the wind and espoused the most debased English habits of compromise and sentimentality.[6]

The quotation comes from a 1968 Banham article entitled 'Revenge of the Picturesque', an attack on Pevsner written with characteristically vigorous polemicism for Pevsner's *festschrift*. Banham had clearly not recovered from the partisan hostilities in which he had played a central role and of which he now wrote – indeed, he says explicitly that 'the battles of the early nineteen-fifties are still being fought'.[7] Perhaps this once-angry controversy is responsible for the vengeful car-tyre marks across the flysheet of the Royal Institute of British Architects (RIBA) Library's reference copy of the book.

The present article revisits Banham's piece now that the fight is over. It will re-examine the controversy through the lens of Adrian Forty's Barthesian definition of architecture as 'a three-part system constituted out of the building, its image (photograph or drawing), and its accompanying critical discourse (whether presented by the architect, client or critic)'.[8] Following this tripartite structure – though not in Forty's order – I will look first at the written evidence on the relations between Townscape and Brutalism. I will then investigate whether the hostile tone of much of this written material is borne out in the imagery of the two movements, and in their built results.

This research takes its cue from Banham, but also from two recent studies of this period in the US and Britain. One, an article by Timothy Rohan on the leading American Brutalist, Paul Rudolph, demonstrates the surprising influence of Gordon Cullen's Townscape sketches on Rudolph's publicity drawings for a Boston office building.[9] The other, Andrew Higgott's *Mediating Modernism,* discusses Townscape and Brutalism together as interrelated manifestations of the same move away from CIAM (Congrès Internationaux d'Architecture Moderne) universalism to an architecture and urbanism of greater local specificity and greater psychological sophistication.[10] The present article aims to share the 'pluralist scepticism' of contemporary historians like Alan Powers, standing outside the debates which for decades were so intense and divisive, and looking at post-war architectural endeavours not as a series of coherent oppositions but as a complex continuum.[11]

Word

Starting, then, with words, to what extent does writing of the post-war period support the notion of a rivalry between Brutalism and Townscape? Banham's 1968 article provides the most complete first-hand account of the controversies of the 1940s–1960s, but his personal involvement in – indeed leadership of – the Brutalist faction in the debate is legible between the lines. His *parti-pris* periodically bursts from the text, ignoring for example the anti-suburban agenda which was central to Townscape and which the Brutalists shared:

> The Picturesque was seen – correctly – as one of the historically contributing causes to the visual disorders of suburbia, and thus fell under the same anathema.[12]

Indeed, in his 1966 book *The New Brutalism: Ethic or Aesthetic,* Banham paints the Picturesque revival as one of the central founding discontents of Brutalism:

> The fundamental command of picturesque theory 'to consult the genius of the place in all' [. . .] seemed to be employed to justify, even sanctify, a willingness to compromise away every 'real' architectural value, to surrender to all that was most provincial and second-rate in British social and intellectual life.[13]

In both the 1968 article and the 1966 book Banham employs the emotive language of 'forgiveness' when recalling the 1940s and 1950s *Architectural Review,* 'whose enthusiasm for picturesque planning has still not been forgiven by some of the Brutalist generation'.[14] Richards's book *The Castles on the Ground,* meanwhile, was 'an apotheosis of English suburbia for which some have never forgiven him'.[15]

The ground over which battle was joined between Brutalists and Townscapists was often the curious one of 'Englishness'.[16] Englishness was perhaps a natural rallying cry in the immediate aftermath of the war. For Pevsner's project of bringing modernism to acceptance by the English it was an obvious tool. Also, Pevsner came from a Germanic academic tradition which emphasized 'national characteristics' – an intellectual structure with uncomfortable echoes of the beliefs which had driven Pevsner out of Germany in the 1930s. The *Architectural Review*'s Picturesque campaign was couched in nostalgic terms as being characteristically 'English'. In his 1955 Reith Lectures, 'The Englishness of English art', Pevsner characterized the Picturesque as England's greatest contribution to world art.[17] He argued in his final lecture that modernist architecture could be planned and landscaped prettily and thoughtfully to achieve a highly agreeable and economical environment. The supreme triumph of this Picturesque site-planning and landscaping movement came at the 1951 South Bank Exhibition of the Festival of Britain. The Exhibition's adoption of Picturesque Townscape as its planning principle was celebrated by extensive, enthusiastic coverage in the *Architectural Review.*[18]

Surprisingly, in view of their self-conscious internationalism, the young hard modernists who gathered around Banham and the Smithsons – a generation largely excluded by their youth from prominence in the Festival of Britain – accepted Englishness as a battleground. Thus Rowe's letter quoted above explains his admiration for Connell, Ward and Lucas, whose 1930s houses 'are so authentic and so English, and yet rise so far above that provincial quality of "Englishness" lately so much valued, that they have still, after all these years, the invigorating qualities of a manifesto'.[19]

Connell and Ward were both New Zealanders, but nevertheless the battle lines were drawn: Pevsner and the *Architectural Review* on one side, arguing that the Picturesque and a sense of accommodating compromise were the true English spirit derived from the part-Welsh Uvedale Price; and Rowe, Banham and the Smithsons on the other, arguing that Englishness was the uncompromising continental-European modernism of the antipodean Connell and Ward.

To these younger architects, design decisions should be justified with reference to theoretical positions rather than through an assumption of universal aesthetic values. A 1954 letter to the *Architectural Review* adds a further component to the requirements of good modernist architecture: it must be didactic. The letter, a response to one of Pevsner's Picturesque articles, is signed by Alan Colquhoun but was, according to Banham, 'the outcome of considerable discussion among the offended parties'.[20] Colquhoun writes:

> Not to admit the didactic element in modern architecture is to make nonsense of it. But once this didactic element is admitted no purely visual 'theory' basing itself on the universal validity of forms independent of structure and function appears to be adequate. It is because this fact has been lost sight of that so much of Post War British architecture is effete and superficial.[21]

Both sides in the fight justified their positions with reference to architectural history. Pevsner was the leading historian and theorist of the Townscape/Picturesque axis, while the

Brutalists favoured John Summerson and in particular Rudolf Wittkower at the Warburg Institute, who had taught Colin Rowe. Wittkower's advocacy of Palladian rationalism mirrored the search of the young for theoretical and formal rigour.[22] Wittkower's *Architectural Principles in the Age of Humanism* emphasized links between leading Renaissance architects and the intellectual avant-garde of their time – a precedent which must have appealed to the intellectual aspirations of the younger generation.[23] The geometrical and mathematical rules stressed by Wittkower's analyses of Palladio were a key generator of the plan of the Smithsons' early manifesto building, Hunstanton Secondary Modern School (1949–54). This formal plan contrasted dramatically with the Festival of Britain's painterly, apparently haphazard organization of spaces and relationships.

In their written style, too, the Brutalists searched for the same aggressive lack of decorative frills and the near-sombre rigour seen in the detailing of Hunstanton. Pevsner, with his magisterial tone and undisguised agenda, neatly filled the role of an authority against whom the younger generation could rebel. Still better were the orotund articles which periodically appeared in the *Architectural Review* under one or another of Hastings's outlandish pseudonyms (Ivor de Wolfe, Ivor de Wofle etc.).[24] Even their titles had a Gilbert and Sullivan quality of mockable and archaizing pomposity: 'Townscape: A plea for an English visual philosophy founded on the true rock of Sir Uvedale Price'.[25] Hastings's writing was studiedly old-fashioned and rich in self-conscious decorative flourishes, as where he complains of:

> the brickish-a-brackish mid-twentieth century world of barbed wire, pig-wire, steel-wire, wire-mesh, telephone wire, electric cables on crazy fir standards, through which as through a cage darkly we are permitted to get an eyeful of lonely villas, poultry farms, Radar stations, motor-car graveyards, Homes for Incurables – all clipt around with plantations of larch and fields of surprised looking wheat. [. . .] The contemporary world is a kind of visual refuse heap, if not insanitary, inelegant, with the shameless utter inelegance of an upset dustbin.[26]

Rowe, the Smithsons and Banham could hardly have asked for a better Aunt Sally to aim at. The contrast with this older modernist 'establishment' heightened the image of the young as radical, contemporary and above all rigorous, against the flowery compromisers of the older generation. Thus the 1956 Connell, Ward and Lucas special issue of the *Architectural Association Journal* which Rowe used as a stick with which to beat the Townscape campaigners was also admired by Peter Smithson:

> On consideration I still think [. . .] that Connell, Ward and Lucas were the nearest we had in England to first generation modern architects.
>
> "High and Over" was built at the same time and in the same spirit as the Weissenhofsiedlung. [. . .][27]

But Peter Smithson went further than Rowe in the search for rigour:

> What C., W., and Lucas lacked was the absolute ruthlessness of Stam or Oud, who could drive the purist aesthetic right through the organisation – into the window detailing, into the furniture, and into the lives of the occupants.

The contrast of rhetoric between this tone – direct almost to the point of brutality – and the rolling prose of Hastings could scarcely be more striking.

Yet the picture is more complex than comparisons of textual style alone might suggest. The *Architectural Review* combined the vigorous and consistent pursuit of its own campaigns with a liberal tendency to give plenty of page space to those who disagreed with it. Major articles by the Smithsons, Banham, Rowe and Stirling appeared in the *Review* in these years, and buildings and projects by this younger generation were covered sympathetically and with plentiful images. The *Architectural Review*'s pluralism saw it publishing everything from Philip Johnson's commentary on Hunstanton School, through Stirling's anxious, confused thoughts on Ronchamp, to their own Townscape articles and a great diversity of historical research, like the two-and-a-half pages of dense text and illustrations in the March 1953 issue, discussing a little-known collection of eighteenth-century architectural drawings held at the RIBA and British Library.[28] Banham confirms that this diversity did lead the young to read the *Review*.[29] Until 1953, indeed, when Theo Crosby became Technical Editor of *Architectural Design,* making it the first choice for the publication of essays and projects for the young avant-garde, the *Architectural Review* was the main vehicle for both sides of the argument. With a magnanimity characteristic of its management they employed Banham, their noisy opponent, on the magazine's editorial staff.[30]

The two sides were, then, in the closest of contact: Banham was Pevsner's PhD student, and they worked together on the *Architectural Review*; the young read the journal for want of alternatives and because their own leaders published in it. It is not astonishing that, for all their noisy proclamations of hostility, there were quiet similarities between these self-proclaimed rival camps.

The most striking of these similarities is a shared hostility to the *tabulae rasae* of pre-war modernist planning as advocated by CIAM. Colquhoun has demonstrated the ingenuity with which Le Corbusier's buildings of the 1930s responded highly specifically to their sites in such a way as to appear that they were non-site-specific, generic solutions slammed into the existing city fabric with respect only for orientation, interrupting 'the continuity of the urban tissue'.[31] The implication of this and of Le Corbusier's numerous unasked-for town planning proposals was that the existing city – insanitary, impractical and irredeemably bourgeois in structure – was to be kept only as a temporary expedient while preparation was made for a modernist new start. This proud hostility to the pre-modern city appears to have subsided among most architects after 1945 (perhaps partly in response to the widespread experience of seeing towns actually being destroyed in the war), and in both Townscape and Brutalist writings one finds explicit opposition to CIAM's destructive tendencies and its preference for generic solutions.

Although the tone of their writing differs, there is clear similarity in content between Thomas Sharp's warning in *Exeter Phoenix* that 'abstract principles of town planning do not in themselves produce a good plan'[32] and Denys Lasdun's statements 'any attempt to produce *a priori* solutions in architecture can only lead to sterility' and 'we must get away from the rigid or overall plan and concentrate on the detailed examination of precise circumstances in precise situations'.[33] Sharp's view that 'the planner must try to catch the personality and character of the place he is planning'[34] was shared by Lasdun when he roamed Bethnal Green in order to get a feel for the terraced streets and to talk to the people to be housed in his council flats there.[35] The Smithsons, in their use of site photos as a backdrop to their Golden Lane competition entry (1951–52), made clear that their building was not a generic solution like a pre-war CIAM town planning statement, but rather a specific building for a specific site. Even their radical Hauptstadt scheme (1957–58) – a scheme so visually unfamiliar that the *Architect and Building News* published one of the sketches upside down – makes a point of overlaying their new network of walkways and roads over the existing city fabric rather than demolishing all of it.[36] This new specificity was one of the central mechanisms of self-definition in the Team X group of young architects who were to play a central role in the termination of CIAM.

If Townscape and Brutalism shared a dislike of what they saw as the reductive rationalism of pre-war planning, their proposed remedies also bore affinities: each sought an architecture and urbanism with which people would engage more fulfillingly at a psychological level. Where Sharp and Townscape proposed to enrich this experience through aesthetic pleasure, the Brutalists tended to frame their reaction in terms drawn from sociology. A Brutalist lexicon of 'community', 'cluster' and 'grain' seems outwardly unrelated to the Townscape terminology of 'enclosure', 'composition' and 'character'. Fundamentally, however, both movements sought urban forms which would engender a greater or more positive psychological response in users and inhabitants.[37]

The differences between the two movements blurred further as the 1950s progressed. With the Smithsons' scheme for Sheffield University in 1953, they and Banham turned from Palladian formality to 'topology' as the controlling rigour for their plans.[38] Taking poetic inspiration from a mathematical discipline for understanding abstracted shapes, they turned it into a formula for producing buildings which purported to have wholly escaped visual criteria of organization. However, it took a subtle mind to adopt this intellectual position, and to the casual observer the shapes of their Sheffield University scheme or its closest built relative, the Park Hill flats (again in Sheffield, by Jack Lynn and Ivor Smith, built 1957–61), were hard to distinguish from shapes freely composed on visual criteria – Picturesque compositions. To an architectural world trained primarily in the visual disciplines of sketching and drawing, the easiest lesson of the two Sheffield projects appears to have been that it was possible to make visually stunning buildings from irregularly angled spine blocks. Denys Lasdun, Andrew Derbyshire and John Weeks used this discovery on green-field sites at the University of East Anglia, Stirling University and Northwick Park Hospital respectively, but the irregularity sanctioned by 'topology' was most valuable on awkward urban sites. Architects could now fill their site to the boundary unapologetically, enjoying rather than disguising the random angles in the perimeter; Paul Rudolph's Government Service Center in Boston shows this tendency at its most expressive.[39] So even as they rebelled against Picturesque planning by pursuing a mathematical idea which hoped to transcend visual planning, the Smithsons and Banham accidentally legitimized a freedom from right-angular geometry which was taken up as an expressively Picturesque tool by their admirers.

Image

If the theories of Townscape and Brutalism show elements of convergence in the 1950s, what of their imagery? The emphasis on the old, the pretty, the villagey in the historical precedents favoured by Sharp could hardly differ more from the sci-fi agenda of high Brutalism, with Banham and the Smithsons praising American car design and the Futurist architectural fantasies of the 1910s.

FIGURE 12.1 Albert Dock, Liverpool, 1956 photograph by Eric de Maré used in the first Townscape casebook; detail of paving emphasizing texture and robustness (Architectural Press archive/RIBA Library Photographs Collection.)

Yet most of the *Architectural Review*'s visual examples of good Townscapes *trouvés* are anything but soft. The illustrations to their first Townscape casebook (Figure 12.1) revel in the tough, dark materials and uneven textures of old industrial streets – an interest shared across the arts, notably in photography and film.[40] The rough, industrial settings of the 1961 film *A Taste of Honey* could belong in the Townscape manuals. Some Brutalists mocked this veneration of industrial heritage. Banham recalls students at the Architectural Association giving satirical gifts of such *objets trouvés* – or *objets volés* – to the editors of the *Architectural Review,* including iron drain covers and chunks crowbarred from the Cobb at Lyme Regis (a harbour wall hailed by the photographer Eric de Maré as 'the Parthenon of the functional movement').[41]

Architectural Review editor J. M. Richards had studied old warehouses, docks and mills since the 1930s, and in 1957 published an issue entitled 'The Functional Tradition', in which he developed further Townscape's rediscovery of the joy of industrial heritage.[42] Whilst its written rhetoric is old-fashioned for 1957, its images show a sensibility remarkably close to that of the Brutalists over the following years. James Stirling, the Smithsons and other young 'hards' were also learning to love Britain's old industrial architecture at this time. Stirling and Gowan's most clearly Brutalist work, the Langham House Close flats, was claimed for the Functional Tradition by the *Architectural Review,* and Stirling showed his interest in the tag in his 1960 article '"The Functional Tradition" and expression'.[43]

The moody, high-contrast black-and-white photography which Richards used to document and dramatize the 'Functional Tradition' was mostly by Eric de Maré, with contributions by Richards and John Piper.[44] Yet their style is similar to that adopted for photographing new Brutalist buildings. The photographs share a similar emphasis on texture; hard surfaces; simple, robust detailing (often over-scaled); and unpopulated or under-populated spaces shot from low down to emphasize their monumental character (Figure 12.2). When people do appear they are romantic waifs whose fragility seems rather to emphasize the solidity of their context than to humanize it.

Outside the world of photography, the imagery of Townscape is most closely associated with Gordon Cullen's sketches. A genius at evoking lively atmospheres through a few bold lines, Cullen produced most of Townscape's analytical and instructional material. Yet he was also asked by architects with impeccably Brutalist credentials to illustrate their schemes – architects including the Smithsons, Lasdun, and Chamberlin, Powell and Bon.[45] While his appeal may have been primarily that he drew so persuasively, it indicates that they were not embarrassed to associate themselves with Townscape's leading visualizer.

The question of image, then, is even more complicatedly interleaved than that of the written word in this supposed clash between Brutalism and Townscape. Townscape's adherents were less soft and humane in their imagery than in their words, revelling in the same sublimity sought by the Brutalists. The Brutalists, meanwhile, enjoyed Townscape's images even as they shouted their hostility to its written agenda.

FIGURE 12.2 Denys Lasdun & Partners, Keeling House 'cluster block', Bethnal Green, 1956–59, photographed by Tom Bell; the emphasis on textures and landscaping is comparable to that of Townscape (RIBA Library Photographs Collection.)

Building

What, then, of my final category, buildings? Do Brutalist buildings bear out the aversion to Townscape ideals proclaimed in their words? The answer is an emphatic 'no'.

Beyond the influence of industrial heritage discussed above, there are at least two significant lessons which Brutalists appear to have learnt from the Townscape recipes published by the *Architectural Review:* attention to landscaping, and Picturesque techniques for producing psychological effects. Moreover, in the final part of this article two case studies will show the extent to which both Lasdun and the Smithsons adopted the softening techniques of Townscape's contextualism for buildings which nevertheless found immediate acceptance into the hard modernist canon.

The first lesson that Brutalist architects absorbed from Townscape, then, was a heightened awareness of the importance of carrying through high-quality design to all aspects of a project, inside and out. Townscape's commitment to the small-scale business of signage, paving, bollards and so on, though attacked by Joseph Rykwert as a preoccupation with the surface at the cost of the underlying organizational structures, made its mark on architects of the 1950s and 1960s.[46] The best Brutalists saw the fundamental importance of keeping tight control of these aspects to maintain the visual force of their buildings.

Brutalist landscaping is frequently an important part of the architecture of a scheme and tends to be noticeably more thought-through than the shadow-gap or *pilotis*-and-grass ground level of many earlier modernist schemes. John Bancroft describes the transition at his Pimlico School between the tarmac of the playground and the striated concrete of the building (Figure 12.3).[47] The two textures blur into each other through the striped concrete and tarmac pathways in between and the small ramp of wall-style concrete which mediates between building and ground. Elsewhere cobbles, a fetish of the 'Functional Tradition' and Townscape, are also a cliché of Brutalist landscaping, appearing (often reused) at buildings as diverse as Chamberlin, Powell and Bon's Golden Lane; Hutcheson, Lock and Monk's Paisley Civic Centre; Gillespie, Kidd and Coia's Cardross Seminary; and Lasdun's Claredale Street housing. This sort of considered landscaping is characteristically Brutalist and archetypally Townscape.

Beyond this attention to (chunky) detail, the Brutalists may also have picked up from Townscape some of their methodologies for producing psychological impact. This interest in the psychological effect of spaces can be seen in proto-Brutalist thought, though without Cullen's casebook empiricism, as early as November 1941, in Ernö Goldfinger's article 'The sensation of space', in which he attempts to categorize the perceptual effects of different shapes and scales of enclosure.[48]

The tentative diagnostic essays of Cullen showed existing examples of pleasurable or exciting urban compositions, generally on a modest scale, made up of every-day buildings from various periods. When, as at the Festival of Britain, these effects were used from scratch, they remained sea-sidey and pretty. The same techniques were later appropriated by the Brutalists as a guide to producing their own far-from-tentative buildings. As projects grew in size, architects ignored Cullen's definition that 'one building is architecture but two buildings are townscape'.[49] Now single vast buildings became powerful Townscapes in themselves, in which one architect had control of all the pathways, vistas and obstructions which in Cullen's and Sharp's historical case studies had generally developed over centuries. Where Townscape had foreseen pleasant, modest buildings interacting to produce gentle satisfaction in the viewer, the 1960s Brutalist search for the spectacular took these recipes and – fusing them with a monumentality

FIGURE 12.3 John Bancroft, GLC Architects' Department, Pimlico School, 1966–70, photograph by Barnabas Calder during demolition, 2008; the ribbed concrete of the building is echoed in the landscaping

evocative of ancient temples, and sometimes a Beaux-Arts processional character sanctioned by Le Corbusier's *promenades architecturales* – tripled the quantities, to stimulate something more like enjoyable terror (Figure 12.4). Though their initial impulse was similar, Rudolph, Lasdun, John Andrews and their fellow Brutalists were Salvator Rosa to Cullen's Claude Lorrain.

If many Brutalists subverted Townscape by exaggerating it into near-Baroque expressiveness, a few of the best also took its plea for contextual sensitivity more seriously than the popular reputation of 1960s architecture would suggest. Lasdun and the Smithsons approached high-profile jobs on sensitive sites with a contextualism which may have been triggered by fear of planning objections but which became a major inspiration for their architectural approaches to the commissions. Two projects completed in 1964 show this, and the next section of this article consists of detailed case studies of these buildings – Lasdun's Royal College of Physicians and the Smithsons' Economist Building. The architects of each re-examined the relationship between the existing city and new buildings added to it, and both projects faced the challenge of being situated not in obscure suburban streets, but on sites of sensitivity and prominence, with powerful neighbours to object if their schemes were seen as disrespectful to their context.

The Royal College of Physicians is in Regent's Park, overlooking the Park and facing buildings by John Nash. It is on Crown Estates land and therefore was subject to their approval as well as the normal planning constraints. Nash's town planning had achieved a new level of respect and admiration in the immediate post-war years through the mediation of John

FIGURE 12.4 Paul Rudolph, Government Service Center, Boston, 1966–71, photograph by Barnabas Calder, 2008; Picturesque composition but Sublime aesthetics

Summerson. Lasdun's task in building an assertive new building on such a sensitive site was a major challenge.[50] Where Le Corbusier might have intentionally affronted the neighbours to imply that they were to be demolished as soon as funds could be found, Lasdun recognized their permanence and value.

How, then, did he try to fit in with them? First, he acknowledged that he needed not only to produce an object building satisfactory in itself but also to consider the external space which it created – an idea going back via Sharp and Sitte to the Baroque, and especially important to Nash's planning of Regent's Park. In early designs Lasdun considered an open-ended court towards the Nash terrace to exploit the intimacy of the cul-de-sac.[51] In the built design he created a similar feeling of enclosure by projecting his lecture theatre from the side of the main building to embrace a garden. The 'medical precinct' which this created was originally a purely architectural idea but has subsequently become an institutional reality through the College's acquisition of leases on the other buildings in St Andrews Place for its own offices and those of related medical organizations.

Second, Lasdun's materials were frankly contextual (Figure 12.5): mainly off-white mosaic tile to match the standard Crown Estates' stucco colour of the Nash. To achieve the match, Lasdun commissioned a specially made colour of porcelain tile from the makers in Candolo, near Turin.[52] Subsequent changes in the stucco colour of the surrounding buildings, now repainted a pinky-yellow cream, have betrayed this effect.

Another colour match was also sought by Lasdun in his design for the Royal College of Physicians: the hipped slate roofs of the Nash terraces suggested to Lasdun the dark blue

FIGURE 12.5 Denys Lasdun & Partners, The Royal College of Physicians, London, 1959–64, photograph by Barnabas Calder, 2012; the off-white tile in the foreground matches the off-white stucco of the nineteenth-century terraces around it, and the black brick of the lecture theatre to the right picks up the colour, texture and shape of the older buildings' roofs.

engineering bricks, and probably the distorted hump of his lecture theatre. Lasdun picked out the two features which he regarded as most strongly and importantly characteristic of Regent's Park – the colour palette of creamy stucco and blue slate, and the planning device of shaping outdoor space using the perimeters of the new building – and with them he accommodated his new building to the *genius loci*.

The Economist Building, also by young hard modernists (Alison and Peter Smithson), was likewise contextually sensitive. The site was located in a conspicuous part of the West End of London. The client wanted a high-rise tower as part of the complex, at the top of which he (the proprietor of the *Economist* magazine) was to live. A plot ratio of 5:1 was a major influence, and the Smithsons chose to divide the various components of the project into three separate towers above a podium.[53] The planning of the piazza which links them was extolled by Cullen in *Architectural Review* as a notable piece of Townscape:

> The interplay between the group and its enclosure is emphasized by the raised level of the piazza which has the effect of excluding much of the neutral streets and vehicles, so giving a direct contact between the Steinberg-like central diagram and its foil.[54]

Perhaps the clearest specimen of Townscape thinking in the project, however, is its relationship to St James's. From along the road the only visible block is self-effacing (Figure 12.6), maintaining the roofline of its neighbours and picking up with accentuated mullions the verticals of its columned Edwardian context. It also makes a gesture of neighbourliness to Boodle's Club next door, reflecting in its storey heights the first-floor *piano nobile* of the Georgian facade. Cullen admired the slit views (Picturesque glimpses) through the entire site on 45-degree angles:

FIGURE 12.6 Alison and Peter Smithson, the Economist Building, 1959–64, photograph by Barnabas Calder, 2012; as Gordon Cullen points out in the *Architectural Review* the building adopts floor and roof levels from its neighbours and echoes in its mullions the verticality of their columns.

> The blocks are so placed that it is possible to see between them throughout the complex; i.e. the piazza never appears as an enclosure of itself but as a space in relation to the outer space. Sometimes the gap is only a few inches wide [. . .] but this is enough, when one is moving, to explain the position.[55]

After the language of Banham and the Smithsons throughout the 1950s, there is something almost shocking about a photo of a Smithson scheme being captioned 'a raised colonnade captures a vista'.[56] However, architects of the 1960s had considerable influence over the publication of their work in the *Architectural Review,* and it seems unlikely that Cullen praised their project in the teeth of their hostility.[57] Rather, it appears that the Smithsons had by 1965 abandoned Banham's hostility to the Picturesque, adding it to their own architectural thinking.

This Townscape contextualism enraged some critics – the young Kenneth Frampton, for instance, bemoaned the failure of the Smithsons to produce 'a hypothetical scheme for the redevelopment of the whole of the east side of St James's from the Economist site down to Pall Mall'.[58] In other words, he regretted precisely that accommodation to the surrounding scene which Cullen so admired in the scheme.

'The triumph of unacknowledged Picturesqueness'

Despite the griping of a few critics, then, the Brutalists were clearly more open to the Townscape movement than has been generally recognized. John Bancroft, architect of the impeccably Brutalist Pimlico School, admired both Cullen and Sharp throughout and respected the Townscape movement.[59] Hastings, meanwhile, published as his last major plea for Townscape a Frankenstein's monster of collaged Brutalist classics, the exemplary city 'Civilia'.[60]

Banham's anti-Picturesque campaign appears to have suffered near-total defeat. His article 'Revenge of the Picturesque' acknowledged it in a bitter and portentous final sentence:

> It [the Smithsons' 1950–51 Coventry Cathedral scheme] was the last formal and geometrical point in history on which the opponents of Picturesque compromise can rely, so total has been the triumph of the unacknowledged Picturesqueness of the Picturesque's avowed enemies.[61]

Perhaps his previous closeness to the Smithsons may have heightened this tone of personal reproach in Banham's writing.

So was the bitterness which Banham ascribes to an entire generation in fact exclusively Banham's bitterness? Were Sharp's Townscape, the *Architectural Review*'s Picturesque and Brutalism all working together against their common enemies: CIAM absolutism, the suburb and visual clutter? It seems not. As Banham says, angry denunciations like that by Thomas Stevens of 'the two main English vices, picturesque muddle-headedness in planning, combined with casual formal confusion', were important to the self-definition of the early Brutalists, even if the architects who signed up to these denunciations were unable to sustain this clear separation from Picturesque techniques in their design approaches.[62]

Equally, despite the tolerance of the *Architectural Review*'s editors, Sharp – a central figure in Townscape – disliked Brutalism. He argued that architects should be self-effacing in deference to the overall Townscape, advocating new buildings 'which show no stylistic tricks at all but which depend for their effect on being clear, well-proportioned and honest'.[63] He hated the attention-grabbing buildings beloved of the Brutalists, attacking even five years after its abandonment Lasdun's project for three tall science towers on the New Museums Site in central Cambridge: 'the whole character of Cambridge, after its slow evolution over 600 years, had been put at the whim of a single determined architect, and has only been saved because he decided to transfer his activities elsewhere'.[64] Lasdun had in fact used Townscape-like techniques in developing his scheme, echoing in the silhouettes of his towers the profile of the nearby Ely Cathedral, and studying with a photo-montage the visual impact of his scheme on the Cambridge skyline. Sharp, however, unmoved by this relationship to Townscape theory, illustrated the scheme not with Lasdun's mock-up but with ones which Sharp himself had commissioned for the planning enquiry, depicting speculative detailing of generic low quality – detailing unimaginable from the drawing boards of Denys Lasdun and Partners. Coming five years after the planning enquiry this seems more than just a consultant's tactic. Rather, taken together with Sharp's other interventions and campaigns, it looks more like a personal war against Brutalism. Brutalism may have been learning from Sharp, but many Brutalist buildings radically altered skylines and streetscapes. Sharp's campaign had sought to preserve the existing character of England's towns, whereas Brutalists aimed actively to improve them. Perhaps Sharp's hostility to these schemes was deeper for their distorted echoes of his own theory and rhetoric.

The hostility could be reciprocal, too. If Lasdun was grateful to Townscape for some of its lessons, he was not sufficiently so to ignore Sharp's attack on him. Angry at the misrepresentation of his architecture in Sharp's mock-ups, Lasdun investigated the possibility of legal action, proposing that 'all copies of the book must be destroyed [. . .] all copies of the book taken out of libraries [. . .] an apology to be published'.[65]

The relationship between Townscape and Brutalism was, then, close but not without friction. Despite their similarities a sense of opposition was important for the self-definition of the two camps. Indeed, their similarities may have increased the rancour. After all, those who really were opposed to the modernist orthodoxy of the 1950s and 1960s, like McMorran and Whitby, were simply ignored by the *Architectural Review* and Banham; hostility was reserved for those who were similar but slightly different.

As for Banham, after his split with Brutalism over its Picturesqueness, he fled to the early High-Tech of Archigram and Price. Even here he was not safe from the revenge of the Picturesque, however. Though they shared a hostility to Townscape, Rowe and Banham had a stormy relationship, and Rowe was later to damn Banham's beloved Archigram as 'Townscape in a space suit'.[66]

Notes

1. The author delivered a paper on the same topic at the 2007 conference 'Le Beau dans la ville' at the Université de Tours, entitled 'Le laid dans la ville: le "townscape" et ses adversaires brutalistes' (publication forthcoming).
2. Letter from Colin Rowe (1957) *Architectural Association Journal,* 72 (January), p. 163.
3. Connell, Ward and Lucas special issue (1956) *Architectural Association Journal,* 72 (November).
4. 'Inside the pub' (1949) special edition of *Architectural Review,* 106 (October); Cruickshank, D. (1991) 'The Bride – spirit of an age', *Architects' Journal* (2 and 9 Jan): 22–7.
5. Sharp, T. (1946) *Exeter Phoenix: A Plan for Rebuilding,* London: Architectural Press.
6. Banham, R. (1968) 'Revenge of the Picturesque: English architectural polemics, 1945–1965', in Summerson, J. (ed.) *Concerning Architecture: Essays on Architectural Writers and Writing Presented to Nikolaus Pevsner,* London: Allen Lane, pp. 265–73.

7. Ibid., p. 273.
8. Forty, A. (2000) *Words and Buildings: A Vocabulary of Modern Architecture,* London: Thames and Hudson, p. 13.
9. Rohan, T. (2007) 'Challenging the curtain wall: Paul Rudolph's Blue Cross and Blue Shield Building', *Journal of the Society of Architectural Historians,* 66 (March): 84–109 (pp. 91–2).
10. Higgott, A. (2007) 'The shift to the specific', Chapter 4 in *Mediating Modernism: Architectural Cultures in Britain,* London: Routledge, pp. 85–116.
11. Powers, A. (2007) *Britain: Modern Architectures in History,* London: Reaktion, p. 7.
12. Banham, 'Revenge of the Picturesque', p. 265.
13. Banham, R. (1966) *The New Brutalism: Ethic or Aesthetic,* London: Architectural Press, p. 13.
14. Ibid., p. 12.
15. Richards, J. M. (1946) *The Castles on the Ground,* London: Architectural Press; Banham, 'Revenge of the Picturesque', p. 265.
16. For a wider discussion of Englishness and modernism, see Whyte, W. (2009) 'The Englishness of English architecture: modernism and the making of a national international style, 1927–1957', *Journal of British Studies,* 48 (April): 441–65.
17. Online at <www.bbc.co.uk/programmes/p00h9llv> (accessed 2011).
18. South Bank Exhibition special issue of *Architectural Review,* 110 (August 1951).
19. Rowe, letter to *Architectural Association Journal,* p. 163.
20. Banham, 'Revenge of the Picturesque', p. 267.
21. Letter from Alan Colquhoun (1954) *Architectural Review,* 116 (July), p. 2.
22. Banham, 'Revenge of the Picturesque', p. 268.
23. Wittkower, R. (1949) *Architectural Principles in the Age of Humanism,* London: Warburg Institute.
24. Lasdun, S. (1985) 'H. de C, the man from the A. P.', *Harpers and Queen International* (April): 210–18 (p. 210).
25. De Wolfe, I. (pseudonym of Hastings, H. de C.) (1949) 'Townscape: A plea for an English visual philosophy founded on the true rock of Sir Uvedale Price', *Architectural Review,* 106 (December): 354–62.
26. Ibid., p. 355.
27. Letter from Peter Smithson (1956) *Architectural Association Journal,* 72 (December), p. 138.
28. Lang, S. (1953) 'History: Visentini drawings', *Architectural Review,* 113 (March): 192–4.
29. Banham, 'Revenge of the Picturesque', p. 267.
30. Lyall, S. (2004) 'Banham, (Peter) Reyner (1922–1988)', in rev. *Oxford Dictionary of National Biography,* Goldman, L. (ed.) Oxford: Oxford University Press; online edition May 2008.
31. Colquhoun, A. (1989) 'The significance of Le Corbusier', in *Modernity and the Classical Tradition: Architectural Essays 1980–1987,* Cambridge, MA: MIT Press, pp. 163–91 (p. 175).
32. Sharp, *Exeter Phoenix,* p. 11.
33. Lasdun, D. and Davies, J. H. V. (published anonymously; in the year-long series of 'Thoughts in progress' dialogues the italicized speaker is Lasdun), 'Thoughts in progress: the New Brutalism' (1957) *Architectural Design,* 27 (April): 111–12 (p. 112); 'Thoughts in progress: the Pavillon Suisse as a seminal building' (1957) *Architectural Design,* 27 (August): 223–4 (p. 224).
34. Sharp, *Exeter Phoenix,* p. 9.
35. 1996–97 interviews by Jill Lever, 'National Life Story Collection: Architects' Lives: Sir Denys Lasdun', online at <http://sounds.bl.uk/View.aspx?item=021M-C0467X0009XX-0100V0.xml> (accessed 2011).
36. Smithson, A. and P. (1958) 'Hauptstadt Berlin', *Architect and Building News,* 214 (2 July): 7–10.
37. Higgott, 'The shift to the specific', p. 110.
38. Stalder, L. (2008) '"New Brutalism", "Topology" and "Image": some remarks on the architectural debates in England around 1950', *Journal of Architecture,* 13: 263–81.
39. Calder, B. (2007) 'Government Service Center, Boston, MA', Twentieth Century Society Building of the Month, July 2007, online at <www.c20society.org.uk/botm/archive/2007/government-service-center-boston-ma.html> (accessed 2011).
40. Cullen, G. (1949) 'Townscape casebook', *Architectural Review,* 106 (December): 363–74.
41. *Architectural Review,* 107 (January 1950), p. 11.
42. Richards, J. M. (1980) *Memoirs of an Unjust Fella,* London: Weidenfeld and Nicolson, p. 194; Richards, J. M. (1957) 'The Functional Tradition', *Architectural Review,* 122 (July): 4–73.
43. Erten, E. (2004) 'Shaping "the second half century": *The Architectural Review, 1947–1971*', unpublished doctoral thesis, MIT, pp. 256–7; Stirling, J. (1960) '"The Functional Tradition" and expression',

Perspecta, 6: 88–97; Vidler, A. (2010) *James Frazer Stirling: Notes from the Archive,* Montreal and New Haven: Canadian Center for Architecture and Yale University Press, pp. 99–103.

44. Elwall, R. (2011) '"How to like everything": Townscape and photography', paper delivered at 'Townscape' symposium, Bartlett School of Architecture, 23 July.
45. For example Cullen's sketches of the Hallfield Estate executed by Drake and Lasdun, RIBA Drawings Collection, PA2095/1(35–36), or Cullen's sketches of the New Museums Site project by Denys Lasdun and Partners, RIBA DR42/6(1). For Chamberlin, Powell and Bon see Harwood, E. (2011) *Chamberlin, Powell and Bon,* London: RIBA. See also Otto Saumarez Smith's study of the influence of the influence of Townscape on Chamberlin, Powell and Bon, in his unpublished master's thesis entitled 'Chamberlin Powell and Bon's Golden Lane Estate and the Counter Attack against Subtopia' (University of Cambridge Department of Architecture). For the Smithsons, see discussion of the Economist Building below.
46. Rykwert, J. (1959) 'Review of a review', *Zodiac,* 4 (Autumn): 13–15; Rykwert spoke at the 2011 Bartlett Townscape symposium and stated that he had not changed his mind in the intervening years.
47. Discussion between author and John Bancroft at Pimlico School, autumn 2007.
48. Goldfinger, E. (1941) 'The sensation of space', *Architectural Review,* 90 (November): 129–31.
49. Higgott, 'The shift to the specific', p. 108.
50. Calder, B. (2008) *A Monumental Act of Faith: Denys Lasdun's Royal College of Physicians,* London: Royal College of Physicians.
51. RIBA Drawings Collection, PB892/1(1–29).
52. 'The Royal College of Physicians, Regent's Park, London' (1965) *Architectural Design,* 35: 274–85 (p. 278).
53. Scalbert, I. (1995) 'Architecture is not made with the brain: the Smithsons & the Economist Building plaza,' *AA Files,* 30 (Autumn): 17–25.
54. Cullen, G. (1965) 'Criticism: the 'Economist' buildings, St. James's', *Architectural Review,* 137 (February): 114–24, p. 123.
55. Ibid., p. 123.
56. Ibid., p. 124.
57. For architects' control of publication in the *Architectural Review* see for example Denys Lasdun's archive at the RIBA Archives Collection. The extent to which he could choose 'star photographs' and so on is visible already by the early 1950s with Hallfield School, RIBA LaD/4/6.
58. Frampton, K. (1965) 'The Economist and the Hauptstadt', *Architectural Design,* 35 (February): 61–2 (p. 61).
59. Discussion between author and John Bancroft at Pimlico School, autumn 2007.
60. De Wolfe, I. (1971) *Civilia: The End of Sub Urban Man: A Challenge to Semidetsia,* London: Architectural Press.
61. Banham, 'Revenge of the Picturesque', p. 273.
62. Stevens, T. (1956) 'Connell, Ward and Lucas, 1927–1939', *Architectural Association Journal* special issue (November): 112–13 (p. 113).
63. Sharp, *Exeter Phoenix,* p. 109.
64. Sharp, T. (1968) *Town and Townscape,* London: Murray, p. 128.
65. Lasdun telecon 25 Nov 1968, memo 26th, RIBA Archives Collection LaD/38/3.
66. Rowe, C. and Koetter, F. (1978) *Collage City,* Cambridge, MA: MIT Press, p. 41.

13

JANE JACOBS, THE TOWNSCAPE MOVEMENT, AND THE EMERGENCE OF CRITICAL URBAN DESIGN

Peter L. Laurence

Introduction

Although she did not come out and say it, some 35 years after she began writing the first draft of *The Death and Life of Great American Cities* (1961), Jane Jacobs alluded to the particular influence of the British Townscape movement on her canonical and subversive book, and on the broader direction of US city planning and design theory. Reflecting on her book's impact on students of architecture and city planning in the foreword to the 1993 Modern Library edition of *Death and Life,* Jacobs wrote that 'their subversion was by no means all my doing. Other authors and researchers – notably William H. Whyte – were also exposing the unworkability and joylessness of anti-city visions. In London, editors and writers of *The Architectural Review* were already up to the same thing in the mid-1950s'.[1]

As observed in the following pages, Jacobs's two modest lines were the briefest summary of a history of ambitions, influences, and collaborations between British and American architectural critics that unfolded in the 1950s, reshaping American architectural criticism, helping to define the new field of urban design, and motivating Jacobs's own influential work.

'Man Made America' and the reinvigoration of American architectural criticism

Although 'Townscape: A Plea for an English Visual Philosophy' was originally articulated by *Architectural Review* editor Hubert de Cronin Hastings in December 1949, as a reaction to developments in postwar Britain, the *Review*'s first major 'Townscape' project focused on the US.[2] 'Man Made America', a special issue of the *Review* published a year later, in December 1950, was a critical case study of 'the mess that is man-made America': warning of what both the US and the UK were becoming; providing a cautionary tale of the political and environmental consequences for Great Britain of American capitalism and consumerism; and making an argument for a critical and more holistic conception of the built environment.[3]

Like other British architects and designers in the early postwar period, the *Review*'s editors – J. M. Richards, Nikolaus Pevsner, Ian McCallum, Osbert Lancaster, and Hubert de Cronin

Hastings (writing under the pseudonyms of 'H. de C.' and 'Ivor de Wolfe') – regarded the US with both admiration and revulsion. On the one hand, they admired aspects of American democracy and the American way of life and felt connected to, even 'personally implicated' in, the outcome of 'the American adventure'.[4] On the other hand, they were repelled in particular by American materialism and the laissez-faire attitude towards the natural and built environment that was turning the US into a 'combination of automobile graveyard, industrial no-man's land, and Usonian Idiot's Delight'.[5] Even worse, the US was exporting its laissez-faire industrialism through the Marshall Plan, which demanded the adoption of American capitalist values in exchange for reconstruction dollars, with direct consequences for the built environment:

> Technocracy, as we see it, is the pistol the U.S. holds to the stomach of western civilization. Though revealing something genuinely heroic in her political handling of the post-war chaos, she is prepared to act big only so long as her fellow-travellers are ready to talk her language. But her language is baby-talk – of dollars and technics – and this is deadly dangerous to democracy. The significance of the American urban landscape is that it exhibits just the same symptoms – the symptoms of infantilism and arrested development.[6]

The *Review*'s editors acknowledged that the UK had already made a mess of its own landscapes, with sprawling, low-density postwar housing and industrial developments that were insensitive to the landscape and of low aesthetic quality. In a backhanded compliment, they offered that Britain's descendants had learned nothing from England's 'visual fate' and acted as if they had no other 'earthly ambition than to provide a bigger, more general suburbia, to add more wire, to model lovingly still huger areas of industrial and even agricultural scabbery'.[7] Despite this harsh criticism, and rather glib self-criticism, the *Review*'s greatest praise, however, lay in the belief that Townscape ideals actually had the best chance of taking hold not in the UK, but in the US. Opining that the US was more receptive to new ideas than the UK, they wrote, 'In the US, if anywhere, its significance might be appreciated'.[8]

Although it took a few years before the *Review*'s ideas about the built environment were appreciated per se, the impact on architectural journalism in the US was immediate. The criticism levelled by 'Man Made America' came as a 'heavy jolt', as Jacobs's boss, *Architectural Forum* editor Douglas Haskell, described it in early 1951. Indeed, not only was Haskell outraged to read such anti-American sentiment coming from a wartime and Cold War ally, but, as a long-time admirer of the *Review,* whose approach he sought to emulate as a writer and editor, he was personally and professionally offended. While *Forum* had in fact loaned photographs to the *Review* to illustrate 'Man Made America', its critique undermined Haskell's long-held romanticism of the American landscape, which – as he had written in an essay published in the *Review* in 1937 – he regarded as both modern and democratic.[9] Altogether, 'Man Made America' went far beyond the architectural criticism that he and other Americans were used to: as Haskell wrote in reply in the April 1951 edition of *Forum,* it was nothing less than a 'wholesale condemnation of American civilization'.[10]

Nevertheless, Haskell found it difficult to make an effective counterargument. He offered various explanations for the nature of the American landscape and offered that there were 'great reservoirs of vitality even in honky-tonk'.[11] But Haskell ultimately had to concur that 'Man Made America' – which featured essays by Christopher Tunnard, Henry-Russell Hitchcock, and Gerhard Kallmann, and photographs and illustrations by Walker Evans, Saul Steinberg, and the *Review*'s Gordon Cullen – had made many accurate and thought-provoking points. He agreed that the US was building a 'supremely ugly . . . tin-can civilization' and acknowledged that 'thoughtful Americans were unreservedly thankful for the sharp reminder, from an outside source that *some* of the "mess" is really there'.[12]

Not coincidentally, approximately a year later, Haskell announced a new editorial policy for *Architectural Forum:* the magazine would take a more aggressive stance in its architectural criticism and increase its focus on urban redevelopment projects. Haskell told his editorial staff in July 1952, which included associate editor Jane Jacobs, who had been added in May, that *Forum* would restore 'genuine architectural criticism – not the wrist slapping kind, but the kind where you first consult your lawyers about possible action'.[13]

Haskell's comment about consulting lawyers was literal. As in Britain, American architectural criticism had been intimidated by the threat of libel suits for some decades. Inspired by the more aggressive stance in England, however, Haskell made a case to his superiors at Time Inc., *Forum*'s parent company, for doing the same. To make his case to *Forum*'s publisher, he alluded to an editorial by J. M. Richards published one month before 'Man Made America', in November 1950, where Richards wrote that because of the libel threat, there was no architectural criticism comparable with that of art, theatre, or music criticism.[14] Paraphrasing Richards, Haskell argued, in late 1951, that 'criticism of architecture should be on par with that of art, music, the theater, and other cultural manifestations'.[15]

It was a challenge for *Forum* to emulate the *Review,* however, and Haskell was sensitive to his magazine's limitations. As he wrote in a long letter defending *Forum* against an unfavourable comparison with the *Review* in 1954, Haskell argued that an American architectural magazine like *Forum* could not be expected to publish architectural theory and criticism of the same level. *Forum,* he explained, had to turn down articles that might be appropriate for the smaller and 'much more literate audience' of the *Review.*[16] Referring to Lang and Pevsner's 1949 article on William Temple's late seventeenth-century concept of asymmetrical composition in the issue that launched the Townscape movement, it was Haskell's conclusion that it was 'for the *Review*'s audience *only* that one could get into great detail about such concepts as *sharawaggi*'.[17]

It also took time for the American situation to catch up with the English, as postwar rebuilding efforts in the UK were necessarily more advanced. Parliament had enacted the Town and Country Planning Act of 1947, dramatically changing the government's control over town planning. By contrast, the US Housing Acts of 1949 and 1954, which provided federal funding for slum clearance and local urban redevelopment projects, remained untested into the late 1950s, insofar as it took many years for the housing and redevelopment projects enabled by these policies to be planned, constructed, occupied, and assessed. Indeed, when Pittsburgh's Gateway Center, one of the first products of urban redevelopment provisions of the 1949 Housing Act, was completed in 1953, American urban theory and architectural criticism remained very immature. While Haskell considered his critique of that project as 'the first piece of architectural criticism that has been so direct and outspoken since around 1928, when two or three magazines retreated in the face of libel suit threats', his editorial was not the scathing attack on urban renewal-era design that might be expected in hindsight.[18] In fact, Haskell's December 1953 editorial – which was illustrated with an image of Le Corbusier's City for Three Million and headlined 'Le Corbusier Made This Prophetic Sketch in 1922 . . . Now, at Last, Office Towers in a Park' – was not an attack on Radiant City-style planning but a celebration of it. What Haskell regarded as a great step forward for architectural criticism was his attack on the project's 'painted on' architecture and the case he made for architects' greater involvement in site planning. Insofar as the architects' responsibilities were generally limited to the decoration of the buildings and not seen to extend to 'plot design', Haskell regarded the success of the Radiant City-like scheme as a fortunate accident.[19]

In other words, in the early to mid 1950s, modernist ideas of the urban figure/ground, context, and historic city fabric were still deeply ingrained. And as Jacobs had observed all of this while she was at *Forum,* the Gateway Center case study (later referred to by Jacobs in *Death and Life*) underscores the revolutionary nature of Jacobs's ideas regarding the integration of building, site, and function, which were inspired in no small part by the *Review* and the Townscape movement.

As Jacobs observed in the quotation at the beginning of this chapter, by 1955 thinking in the US had begun to change. By then, there was some evidence of the failure of prevailing urban theory. Moreover, in June 1955, the *Review* published a UK version of 'Man Made America' in the form of Ian Nairn 'Outrage' special issue (which was followed in December 1956 by 'Counter-Attack against Subtopia').[20] Impressed and inspired by this direction, in late 1955 Haskell wrote again that the *Review* had been correct about 'Man Made America' and agreed that it was time for Americans to think carefully about 'Roadtown' – what the landscape of the US had become.[21] *Forum,* he wrote, would promote a more urban-minded direction for criticism and embrace collaboration with the *Review* in a critical approach to the larger built environment. Preparing the ground for Jacobs's work with Nairn and Cullen a few years later, Haskell wrote:

> Two paths are open to us. One is to accept Roadtown as a formidable fact and civilize Roadtown, now that it is commanding heavier highway engineering and bigger building capital. The other is to re-examine the very roots of our endlessly shuttling civilization. On both these subjects *Forum* will gladly work with the *Review.*[22]

Jane Jacobs and the Townscape movement

In 1956 Jane Jacobs listened to José Luis Sert make his case for the new field of urban design at the First Harvard Urban Design Conference, an event that helped reveal to her the need for *The Death and Life of Great American Cities* and what she described as a 'different system of thought about the great city'. However, as someone who began her work as an architectural critic in the early 1950s with an idealization of the field of city planning, faith in urban renewal, and belief in prevailing city planning theory, Jacobs's ideas about the life and design of cities continued to develop in parallel with, and with awareness of, the Townscape movement and the field of urban design.

As already observed, Jacobs's tenure at *Architectural Forum* coincided closely with the development of Haskell's more aggressive editorial policy, the journal's coverage of planned and built early urban renewal projects, and exchanges with their counterparts at the *Review.* Therefore, it should not be surprising that Nairn 'Outrage' – which Jacobs cited in *Death and Life* – not only prompted Haskell's December 1955 editorial, 'Can Roadtown Be Damned?' – with which he welcomed collaboration with the *Review* – but motivated a similarly significant effort by *Forum,* organized and edited by Jacobs.

Forum's 'What City Pattern?' issue, published in September 1956, included an introduction and contributions by Jacobs, Peter Blake, and other *Forum* editors; guest contributions by

Catherine Bauer and Victor Gruen; and a concluding editorial by Haskell. 'Architecture for the Next Twenty Years', Haskell's editorial, was nominally addressed to the American Institute of Architects on the occasion of the organization's centenary, but referred to 'Man Made America' as its inspiration:

> Back in 1950, friends of ours across the Atlantic, editing England's *Architectural Review,* cut deeply into our native pride with a complete issue devoted to 'Man-made America.' What they said still rankles – because there was some justice in it. They were talking not about our selected beauty spots, our fine suburban homes, our recognized monuments. They were talking about America as a whole . . .[23]

For Jacobs, *Forum*'s 'big planning issue', as Haskell called it, was an important and influential assignment. With it, her idealization of the field of city planning and faith in urban renewal were replaced by a critique of prevailing city planning theory.[24] Catherine Bauer's contribution to the feature was particularly significant in this regard. In it Bauer outlined a comprehensive rejection of prevailing city planning theory as anti-urban, utopian, and unworkable. Despite the notoriety and praise that Jacobs later received for the critical introductory chapter of *Death and Life,* Bauer's critique was even more expansive: whereas Jacobs famously attacked the 'Radiant Garden City' model, Bauer criticized not only Ebenezer Howard's Garden City and Le Corbusier's Radiant City, but Frank Lloyd Wright's Broadacre City and Buckminster Fuller's 'nomadic noncity' as well.[25] Upon reading the first draft of the essay in May 1956, Jacobs reported her enthusiasm for Bauer's new direction for urban theory to Haskell:

> I think this kind of article itself represents a turning point . . . As long as the great planning ideas, both inside and outside the city, have been stimulated and intellectually fertilized by city-rejectors, as they have been, how could less imaginative planners and the unimaginative body of citizenry help but take their cue? What and who was there to lead them in any other direction? In this article of Catherine Bauer's is the start of a new direction and I think it is very exciting.[26]

For Jacobs, 'What City Pattern?' also represented a moment of decision between focusing on the subject and problems of sprawl and the larger built environment, and focusing on the city. Her contributions to this issue (and others) dealt with both subjects. In her introduction to the feature, she presciently summarized one of the greatest American problems, then and since:

> The US is heading into a growth crisis, the like of which was never seen before. It is an unprecedented crisis simply because we are an unprecedented nation of centaurs. Our automobile population is rising about as fast as our human population and promises to continue for another generation . . . And because asphalt will not grow potatoes, the pavement that will be demanded by two cars for every one that we have today will have to come out of [our] other-purpose acreage. There's the rub. For the car is not only a monstrous land-eater itself: it abets that other insatiable land-eater – endless, strung-out suburbanization.[27]

By contrast, another of Jacobs's contributions to the 'What City Pattern?' feature, 'Central City: Concentration vs. Congestion?', outlined her future focus. In this short essay, she articulated a critical tenet of her own system of thought about the great city – the virtue of concentration:

> The very essence of the city is intense concentration of people and activities. For concentration means exchange, competition, convenience, multiplicity of choice, swift cross-fertilization of ideas, and variety of demand and whim to stimulate variety of skill and will. Those are the strengths of the city. The suburbs may be incubators of people, but the city stands supreme as the incubator of enterprises.[28]

It was along these lines that Jacobs parted company and found allies. She was ultimately disappointed with Bauer's 'decentrist' position, an inclination toward greenfield New Towns and faith in what Jacobs regarded as the quixotic cause of regional planning. She did not believe that the US had the political apparatus or economic leverage for effective regional planning. She observed, 'In the United States nothing gets done until the situation is desperate; only because the central city situation is desperate does anything get done about it now and we have the instruments [such as urban renewal legislation]'.[29] Townscape theorists, however, had recognized the significance of the city centre. As Nairn wrote in his book *Outrage* in 1956, under the heading 'anti-urbanism':

> Urban sprawl has come to its second stage: with everyone gone to the suburbs, the centre has been left to decay. Towns have become half alive: one half is where you work, but can't live, the other half is where you live but don't work. Half alive towns will produce half alive people, and the most immediate result is that in between working and living there can be up to two hours of limbo, nearly fifteen per cent of one's waking hours: forced and frustrating comradeship in public transport or forced and frustrating isolation in private cars.[30]

Jacobs also appreciated the Townscape critique of suburbia and the suburbanization of public urban space: the islanded, railed off, and meanly decorated public parks, garlanded with herbaceous borders that reflected 'humdrum suburban life' and demeaned both town and country.[31] She described this in *Death and Life* on numerous occasions as the 'great blight of dullness'.

More significantly, Jacobs embraced the Townscape principle of 'multiple use', the antithesis of segregated land-use zoning, to the extent that she became completely associated with the concept. Although Jacobs was familiar with the vitality of pre-functionalist zoning in industrial Manhattan in the 1930s, Townscape principles reinforced her own latent understanding. In *Death and Life,* she would repeat the distinction between complexity and chaos articulated here in *Outrage:*

> Industry and housing, commercial traffic and pedestrian square, cranes and trees, pub and warehouse, all superimposed, not segregated into zones areas – 'residential, 'industrial', 'recreational', and so on. When put together they interact to give virility, not chaos.[32]

By 1957, a year after 'What City Pattern?', Jacobs had pulled together ideas for what Haskell later described as 'the first comprehensive piece' on the subject of urban redevelopment, and a prelude to *The Death and Life of Great American Cities.* In November 1957, Haskell described the outline of Jacobs's proposed feature to his editorial colleagues and supervisor Ralph 'Del' Paine (publisher of *Forum* and *Fortune*), Joe Hazen (managing editor of *Forum* and *House & Home*), and Lawrence P. Lessing (assistant managing editor of *Fortune* and science writer). Jacobs, he told them, 'has been talking about an approach to city pattern which I think we should discuss very seriously with her because it just might make an impression in *Forum* as strong as our September 1956 ['City Pattern'] issue'.[33] Jacobs was prepared to attack the prevailing concept of 'super-block' development, with its large-scale and top-down planning approach, and to make the case for a fine-grained and intimately scaled city fabric that was sympathetic with Townscape theory. She would argue for more streets instead of fewer, and a greater number of smaller public spaces over a smaller number of large ones. A few weeks later, she was given the go-ahead on her 'blockbuster on the superblock'.

To Haskell's disappointment, Time Inc.'s executive editors decided to divert Jacobs's feature, published as 'Downtown Is for People', to the April 1958 issue of *Forum*'s sister publication, *Fortune,* where it would have more space and greater exposure and would become the capstone of an on-going series of articles on cities and urban sprawl being organized by William 'Holly' Whyte.[34] They hoped to repeat the success that Whyte, an assistant managing editor at *Fortune,* had recently achieved with another series – a sequence of interviews with corporate executives that led to his best-selling book *The Organization Man* (1956) – a study that had stimulated Whyte's interest in the sociology of the suburbs, a subject broached at the same time in 'Outrage' and 'Counter-Attack against Subtopia'.

Although the exact circumstances of the collaboration are unclear, 'Downtown Is for People' was published with heavily edited companion pieces by Ian Nairn and Grady Clay, associate editor of *Landscape Architecture Quarterly* and the *Louisville Courier-Journal*'s real estate editor. It is probable that their contributions were based on Haskell's or Jacobs's recommendations to Whyte. Indeed, although Whyte later claimed that he was the sole organizer of the 'Exploding Metropolis' series, a letter from Haskell to Nairn in May 1958 indicates that Whyte was their 'student', not the other way around.[35] Clay had also been following the Townscape movement for some years on his own; in 1955 he had started a monthly 'Townscape' column in the *Arts in Louisville Gazette* that in 1958 he shifted to the *Louisvillian* magazine, where it received notice around the country.

What is clear is that *Fortune* commissioned a tour of six US cities by Nairn and a tour of 11 cities by Clay. In the course of these travels, Nairn met with Jacobs and Clay: Jacobs hosted Nairn during his visit to New York, and Nairn visited Clay in Louisville, before going on to San Antonio, San Francisco, Chicago, and Boston. He took photographs later sketched by Gordon Cullen that illustrated 'Downtown Is for People' as a mini Townscape 'casebook', which stood alone in the magazine as a sidebar feature titled 'The Scale of the City', emphasizing the pedestrian's serial vision and phenomenological experience of the city.[36] Clay's feature, 'What Makes a Good Square Good?', meanwhile focused on successful public urban spaces in ways similar to those Jacobs would use in *Death and Life.*

'Downtown Is for People' was a tremendous success, and important for all of the contributors' careers. Whyte reported to Haskell that the responses it received were among the most positive of any article published by *Fortune,* with letters of praise from mayors, city planning directors and academics, real estate developers, and urban renewal consultants.[37] It also drew the attention of the Rockefeller Foundation, leading to the 1958 grant that allowed Jacobs to write *Death and Life* and bringing her into contact with others developing alternative approaches to city planning and design. Among these were Kevin Lynch; Christopher Tunnard, who received a Foundation grant in 1957 that led to *Man-Made America: Chaos or Control?* (1963), which harked back to the eponymous issue of the *Review;* and Ian McHarg, who received a Foundation grant for basic research in the field of landscape architecture following a confidential but favourable review of his proposal by Jacobs.[38]

Indeed, in the late 1950s, Jacobs, Nairn, Clay, Cullen, and McHarg were in frequent communication. Following a February 1958 article in the *Louisville Courier-Journal* about Nairn's visit to Louisville, Clay wrote again about the *Review*'s influence in the April 1958 edition of *Landscape Architecture Quarterly.* In an editorial titled 'Critics Wanted! New American Landscape Needs Criticism', he argued that 'subtopia' and 'widespread uglification need the widest variety of criticism'.[39] In mid 1958, he also visited Nairn in London and pitched an article on the *Review*'s 'COUNTERATTACK bureau' to a number of US magazine editors.[40]

Meanwhile, at the University of Pennsylvania, McHarg's proposal for a comprehensive reconsideration of the field of landscape architecture for the postwar world was particularly influenced by Townscape theory, which, following 'Downtown Is for People', McHarg associated with Jacobs and Clay:

> This term ['Townscape'] describes the human view of the city, the consciously and unconsciously created vistas and prospects, of buildings, spaces and buildings, street trees, street furniture, fountains, benches, sculpture, textures of materials, lighting, changes of level, the myriad of smaller constituents of visual experience in the city. This concern is reflected in the recent issues of *Fortune* magazine devoted to the city, and it is reflected in material within a wide range of architectural magazines but it is most closely associated with *The Architectural Review.* This preoccupation is represented in the point of view of Jane Jacobs, Lewis Mumford, and Grady Clay. It invites a concern for the sequence of intimate visual experiences as a determinate of the design rather than emphasis upon the plan, the air view, the rational system.[41]

Later in 1958, Jacobs, Clay, McHarg, and Lynch all met at the Penn-Rockefeller Conference on Urban Design Criticism, which emerged out of the same conversations between Jacobs and Rockefeller Foundation associate director Chadbourne Gilpatric as her grant to write *The Death and Life of Great American Cities.*[42] Although he did not attend the conference, Nairn later received a Foundation grant, again with Jacobs's encouragement, that allowed him to write a sympathetic work, *The American Landscape: A Critical View* (1965).

Ultimately, however, Jacobs's seminal book was perhaps the most important outcome of the conference and the Foundation's larger urban design initiative: her book exemplified the objectives set out for that academic conference – 'first, to improve the quality of critical writing and the philosophy of urban design, and second, to expand the volume of such writing directed toward popular understanding and appreciation'.[43] By virtue of the *Review*'s influence on *Forum*'s editorial agenda, on Jacobs, and on American architectural criticism at large, her book can be also counted as among the most significant outcomes of 'Man Made America' and the broader Townscape movement.[44]

Townscape and urban design

The British Townscape movement and American urban design both grew out of similar reconstruction and modernization needs of postwar cities, and both were reinvigorations of civic design traditions dating back to the 1910s and 1920s, and further back through the Civic Art movement. As Hubert de Cronin Hastings acknowledged in his December 1949 essay 'Townscape', the new philosophy of 'town planning as a visual art' was close to what others called 'Civic Design'.[45] Both movements were also straightforward practices at heart. Gordon Cullen defined 'Townscape' in 1953 by stating that 'one building is architecture but two buildings is Townscape . . . Multiply this to the size of a town and you have the art of environment'.[46] The definition applied equally well to civic design and urban design.

Despite its emphasis on 'visual planning', what initially distinguished Townscape from urban design, however, was theoretical content: a sensibility about the unique phenomena of place; the material, social, and temporal complexity of the city; and the democratic ideals of the inclusivity of everyday urban life. As Hastings observed in his early 'Plea for an English Visual Philosophy', city design required, 'as in politics, a radical idea of the meaning of parts'.[47] This critical aspect of Townscape appealed greatly to Jacobs. She shared Hastings's radical belief that the city planner should, in his words, 'love, or try to love', the diverse expressions and forms of the democratic landscape, 'instead of trying to hate and rid yourself of them in one way or another'.[48] The city, as Hastings described it, was a 'vast field of anonymous design and unacknowledged pattern which still lies entirely outside the terms of reference of office town-planning routine'. Jacobs agreed. As she wrote in 1955, at the time when her hopes that city planning could save the city began to crumble, '[h]undreds of thousands of people with hundreds of thousands of plans and purposes built the city and only they will rebuild the city'.[49]

Thus, although it may seem ironic that a special issue on 'Man Made America' helped put on the map a particularly 'English' landscape philosophy – Townscape theory – its influence on Jacobs, amongst others, brought to fruition Hastings's hope that its ideas would be embraced in the US. While others gravitated towards, and critiqued, Townscape's emphasis on the visual, the movement's critical dimension had a profound influence on Jacobs's work and on American architectural criticism and urban theory at large.

Acknowledgements

Many thanks to Erdem Erten and the editors for their invitation to contribute to this collection, their patience, and their suggestions. More thanks to the Avery Architectural & Fine Arts Library, Columbia University (AAFAL); the University of Pennsylvania Architectural Archives (UPAA); the Rockefeller Archive Center (RAC); Loeb Library Special Collections (LLSC), Harvard Graduate School of Design; Grady Clay and Judith McCandless; David Seamon; Robert Wojtowicz; David Brownlee; and David Leatherbarrow.

Notes

1. Jacobs, J. (1993) *The Death and Life of Great American Cities,* New York: Modern Library, p.xiii.
2. De Wolfe, I. [Hastings, H. de C.] (1949) 'Townscape: A Plea for an English Visual Philosophy', *Architectural Review,* 106 (Dec.): 355–62.
3. The Editors [Richards, J. M., Pevsner, N., McCallum, I., Lancaster, O., and Hastings, H. de C.] (1950) 'Man Made America: A Special Number of the *Architectural Review*', *Architectural Review,* 108 (Dec.), p. 337.
4. Ibid., p. 340.
5. Ibid.
6. Ibid., p. 416.
7. Ibid., p. 415.

8. Ibid., p. 341.
9. Haskell, D. (1937) 'Architecture on Routes U.S. 40 and 66', *Architectural Review,* 81 (Mar.), p. 105.
10. Haskell, D. (1951) 'A Reply to: Man Made America, Special Number of the *Architectural Review*', *Architectural Forum,* 94 (Apr.), p. 158.
11. Ibid., p. 159.
12. Ibid., p. 158.
13. Haskell, D. (1952) Memo to Staff, July 23, 1952, Haskell Papers 57:3, Avery Architectural & Fine Arts Library (AAFAL), Columbia University, New York.
14. Richards, J. M. (1950) 'Architect, Critic and Public', *Royal Architectural Institute of Canada Journal,* 27 (Nov.), p. 372.
15. Haskell, D. (1951) Memo to J. Hazen, Nov. 16, 1951, Haskell Papers 80:8, AAFAL.
16. Haskell, D. (1954) Letter to J. Rannells, Feb. 26, 1954, Haskell Papers 38:1, AAFAL.
17. Emphasis added. Lang, S. and Pevsner, N. (1949) 'Sir William Temple and Sharawaggi', *Architectural Review,* 106 (Dec.): 391–93. Erdem Erten discussed the relationship of sharawaggi and Townscape in 'Shaping "the Second Half Century": The *Architectural Review* 1947–1971', unpublished PhD thesis, Cambridge, MA: MIT, pp. 34–40.
18. Haskell, D. (1954) Letter to W. Wurster, Jan. 14, 1954, Haskell Papers 24:6, AAFAL.
19. See the following by Haskell: (1950) 'Pittsburgh and the Architect's Problem', *Architectural Forum,* 93 (Sept.), p. 29; (1953) 'The Need for Better Planning, and How to Get It', *Architectural Forum,* 98 (June): 146–55, 180–220; (1953) 'Architecture: Stepchild or Fashioner of Cities?', *Architectural Forum,* 99 (Dec.), p. 117; (1953) 'Gateway Center: Now, at Last, Office Towers in a Park', *Architectural Forum,* 99 (Dec.): 112–16.
20. Nairn, I. (1955) 'Outrage (in the Name of Public Authority)', special issue of *Architectural Review,* 117 (June); Nairn, I. (1956) 'Counter-Attack against Subtopia', special issue of *Architectural Review,* 120 (Dec.).
21. Haskell, D. (1955) 'Can Roadtown Be Damned?', *Architectural Forum,* 103 (Dec.), p. 164.
22. Ibid.
23. Haskell, D. (1956) 'Architecture for the Next Twenty Years', *Architectural Forum,* 105 (Sept.), p. 164.
24. For more on Jacobs's early idealization of city planning and favourable writing on urban redevelopment and renewal, see Laurence, P.L. (2011) 'The Unknown Jane Jacobs: Geographer, Propagandist, City Planning Idealist', in Page, M. and Mennell, T. (eds) *Reconsidering Jane Jacobs,* Chicago: Planners Press, pp. 15–36.
25. Bauer, C. (1956) 'First Job: Control New-City Sprawl', *Architectural Forum,* 105 (Sept.): 106–11. Jacobs edited this essay and probably selected the images and wrote the editorial captions.
26. Jacobs, J. (1956) Memo to Douglas Haskell, May 28, 1956, Haskell Papers 24:6, AAFAL.
27. Jacobs, J. (1956) 'By 1976 What City Pattern?', *Architectural Forum,* 105 (Sept.), p. 103.
28. Jacobs, J. (1956) 'Central City: Concentration vs. Congestion?', *Architectural Forum,* 105 (Sept.), p. 115.
29. Haskell, D. (1956) Letter to C.B. Wurster, July 13, 1956, Haskell Papers 24:6, AAFAL.
30. Nairn, I. (1956) *Outrage,* London: Knapp, Drewett & Sons, p. 381.
31. Ibid., p. 386.
32. Ibid., p. 435.
33. Haskell, D. (1957) Memo to J. Hazen, L. Lessing, P. Grotz, D. Paine, Nov. 21, 1957, Haskell Papers 24:6, AAFAL.
34. The circumstances of Jacobs's contribution to Whyte's 'Exploding Metropolis' series were later confused by his 1992 introduction to the revised edition of the book, wherein Whyte overstated his independent editorial influence over the original *Fortune* magazine series and over Jacobs. There he also erroneously reported that Jacobs's work at *Architectural Forum* 'consisted mainly of writing captions' and that she had 'never written anything longer than a few paragraphs' before 'Downtown Is for People'. See *The Exploding Metropolis,* 1993, Berkeley: University of California Press, p.xv.
35. Haskell, D. (1958) Letter to I. Nairn, May 7, 1958, Haskell Papers 2:3, AAFAL.
36. Orillard, C. (2009) 'Tracing Urban Design's "Townscape" Origins: Some Relationships between British Editorial Policy and an American Academic Field in the 1950s', *Urban History,* 36 (2): 294–95.
37. Whyte, W.H. and Kammler, R. (1958) Selection of letters received in March and April 1958 by *Fortune* magazine Letters Dept. re. 'Downtown Is for People' by J. Jacobs, RF RG 1.2 Ser. 200R, Box 390, Folder 3380, RAC.

38. See also Laurence, P. (2006) 'The Death and Life of Urban Design: Jane Jacobs, the Rockefeller Foundation and the New Research in Urbanism 1955–1965', *Journal of Urban Design,* 11 (June): 145–72.
39. Clay, G. (1958) 'Critics Wanted! An Editorial: New American Landscape Needs Criticism', *Landscape Architecture Quarterly* (Apr.), p. 143.
40. Clay, G. (1958) Letter to I. Nairn, May 30, 1958, Personal Papers of Grady Clay, courtesy of G. Clay.
41. McHarg, I. (1958) Research proposal notes under the heading 'Townscape', Ian McHarg Papers [not catalogued], University of Pennsylvania Architectural Archives, Philadelphia.
42. Laurence, 'Death and Life of Urban Design', p. 163.
43. Wheaton, W. and Crane, D. (1958) 'A Proposal to the Rockefeller Foundation for a Conference on Criticism in Urban Design', (12 June), RF RG 1.2 University of Pennsylvania-Community Planning, series 200, box 457, file 3904, RAC.
44. Erdem Erten makes note of an exchange of admiration between Hastings and Jacobs, in the form of Ivor de Wolfe's review of *Death and Life,* 'Death and Life of Great American Citizens', published in the *Review* in February 1963; and in the form of a letter of admiration from Jacobs to 'Ivor de Wolfe' in March 1964. See Erten, 'Shaping "the Second Half Century"', pp. 100–02.
45. De Wolfe, 'Townscape: A Plea for an English Visual Philosophy', p. 362.
46. Cullen, G. (1953) 'Prairie Planning in the New Towns', *Architectural Review,* 114 (July), p. 33.
47. De Wolfe, 'Townscape: A Plea for an English Visual Philosophy', p. 361.
48. Ibid., p. 354.
49. Jacobs, J. (1955) 'Philadelphia's Redevelopment: A Progress Report', *Architectural Review,* 103 (July), p. 118.

14

NEO-REALISM

Urban form and *La Dolce Vita* in post-war Italy 1945–75

Eamonn Canniffe

Introduction

Following the Second World War, damage to Italian urban centres meant that their reconstruction was a priority. The level of destruction in cities such as Florence led to renewed research into the applicability of historical models, free of the propagandistic purposes to which urban heritage had been put under fascism. It was recognized that besides their cultural value, carefully restored and promoted historical environments could offer important tourism potential. Interest in, and observation of, the Italian context was divided between those who concentrated on appearance and those who claimed to see the underlying structure. A more thorough appreciation was claimed by those who claimed to see beyond both the universal models favoured by the older generation of modernists and the intuitive approach of townscape. Team 10, whose analysis of urban form was influenced by contemporary anthropological research, explored the organic nature of traditional environments, which they upheld as models for a more authentic life than was available in the contemporary planned world. In turn, Neo-Rationalism would also look to the morphology of the traditional city for a way forward.

From 1945 the international reception of the Italian urban scene was conditioned by the success of Neo-Realist cinema, with its use of non-professional actors, actual locations and narratives of social injustice. As a movement, Neo-Realism repositioned Italian culture and had parallel manifestations in architecture and urbanism, because it was so closely linked with the urban experience.[1] The films, which had both popular and critical success, portrayed aspects of Roman life during the struggle for liberation and its economic aftermath, by integrating documentary techniques with the use of non-actors, to reflect experience back to the audience, rather than purvey a distracting spectacle. A few seconds of covertly filmed footage of German stormtroopers in the Piazza di Spagna served as the establishing shot for Roberto Rossellini's *Rome, Open City* (1945). That alien presence, occupying a familiar Italian townscape, encapsulated the changes which occurred in Italian urbanism around this date. Later scenes showed the audience other urban landmarks to contextualize the drama, such as the paired towers of Sta Trinita dei Monti and, in the closing frames, the dome of St Peter's Basilica; but this epitome of Neo-Realism otherwise eschews the monuments of the historic centre in favour of the anonymous periphery, where ordinary people lived. With their depiction of the direct brutalities of fascism and the deprivations of post-war unemployment, Neo-Realist films presented a politically engaged position, but they were also commercial products which responded to the popular mood. Traditional urban space was lost from view, except when it was used to underscore contemporary miseries, a disappearance paralleled in the field of urbanism by the architectural concentration on the formal design of housing.[2] There, in the *borgate,* the districts of modern apartment blocks, a seemingly authentic experience was portrayed as the milieu of an heroic popular struggle against the brutal exercise of power, and the traditional language of the historic city was seen as an irrelevance. This turn towards the ordinary and away from the rhetorically representational exemplified the architectural production of the early years of the post-war period, although, as will be discussed later, with the economic recovery this asceticism would stand in contrast against often fabulous visions of modernity and luxury which became available to larger sections of the population.

Political context

In discussing these transformations, however, it is necessary to rehearse the specific political context. Italy's entry into the Second World War in 1940, her surrender to the Allies in 1943 following invasion, the subsequent German occupation, civil war and eventual liberation in 1945 punctured the deluded self-confidence of the cultural elite. Mussolini's adventurism had destroyed the economy and left large numbers of the surviving population suffering disease and in penury. Reconstruction was urgent because of the tense geopolitics of the start of the Cold War. As a frontline state in the new ideological conflict, bordering both Communist-controlled Yugoslavia and Soviet-occupied Austria, Italy benefited from American aid as her industrial base was re-established and the new republican constitution was designed. The experience of civil war had produced a highly fractured political order that would be exacerbated by the onset of superpower tensions between the former Allies. The elections of 1948 were dominated by divisions between former comrades in the resistance and the spectre of Soviet domination through the agency of the genuinely popular Italian Communist Party. The election was won, however, by the Christian Democrats (backed by the Catholic Church and the United States), and they were to remain an almost permanent feature of Italy's numerous short-lived post-war governments over the next five decades. Fascist patronage had compromised all construction over two decades, and architects and planners therefore had to start afresh in their proposals for the reconstruction. The former regime had not only been associated with ponderously rhetorical public work but also with Italian modernism, and the leading figures in the profession expressed a degree of ambiguity in their recreation of architectural languages after 1945, indulging in a form of amnesia which attempted to cleanse pre-war Italian modernism of fascist associations which had now become unpalatable. The manipulative force with which fascism had exploited historical forms for party propaganda, dispensing with its early adherence to modernity, meant that in the initial post-war period any attachment to historical forms would be regarded as politically suspicious. Therefore, this period of post-war reconstruction provided a pause for reflection in the matter of urban space, an understandable reaction to the fascists' use of the rhetoric of the public realm. The political diversity of the new era would occupy these same spaces with a noisy visual battle of rival party posters – the parties' proclamations increased in number and impact during elections and referenda – competing for attention with commercial images and illuminated signage, as well as the more local, customary and transitory black-bordered death notices. The evening *passegiata,* the collective stroll through streets and squares, would be relaxing only if one ignored the cacophonous rival claims to party or brand loyalty.[3]

With the scale of destruction in many European cities apparently presenting a form of tabula rasa to urban designers, Italy's situation in the architectural field was something of a paradox. There had been massive bombardment mainly at the behest of the Allies, with the Germans attempting to salvage a reputation for *kultur* by minimizing the loss of historic environments. In Florence, where Giovanni Michelucci (1891–1990) supervised the reconstruction of the city quarters adjacent to each end of the Ponte Vecchio (Figure 14.1), which had been spared by the retreating Germans, a hybrid medievalized modernism was employed, and eventually significant schools of urban morphology and conservation would spring from this research. But the most immediately pressing matter was the need to improve the living conditions of the growing ranks of the urban poor. This would lead architects and planning authorities to concentrate on providing housing, the best examples of which combined innovative and vernacular technologies to produce new types of popular environments, labelled subsequently with the term Neo-Realist, adopted from films and literature.[4]

The most famous products were the films exposing Italy's reduced circumstances – initially *Rome, Open City* and then other international successes such as *The Bicycle Thieves* (1948) by Vittorio de Sica.[5] The latter film placed its father and son protagonists in the desolate spaces of the modern city, peripheral and poverty-stricken, lacking the consolations of community that a traditional environment might have afforded. However, in a particularly telling sequence shot in the typical streets of the *centro storico,* that sense of community was used to thwart the attempted stealing of a desperately needed bicycle. In both filmic and urban terms this critical vision portrayed modernity as synonymous with want, as part of a project which emphasized the provision for existence but little else. To enable his family's survival, the emblematic man is reduced to dependency on the possession of a mechanical object, and a basic one at that; but this was a phenomenon which pervaded all aspects of Italian life following the collapse of the dominant ideology.[6]

Post-war reconstruction

In the architectural sphere, one legacy of the civil war and the post-war settlement would be a form of clientelism in which the major political parties – the Christian Democrats, the Socialists and the Italian Communist Party (the biggest in western Europe) – directly or indirectly controlled different aspects of the construction market, commercial office development, public buildings and housing, and bribes were expected in return for the allocation of contracts. Entrance to the professions also depended on an academic system where positions in schools of architecture were allocated on the basis of nepotistic connections and political allegiances. This also sometimes manifested itself in the ideological confrontations that characterized architectural debate, reflected in the editorship of the principal journals and reviews. Early in the post-war period these political debates were received as attempts to create a divide between organic and functional architecture within the modern movement itself. Few architecturally significant major projects date from this post-war period, although one major example would be the square and facade of the Stazione Termini in Rome, completed in time to receive the huge numbers of pilgrims for the Holy Year of 1950. The original project was implemented

FIGURE 14.1 Oltrarno, Florence. A detail of Giuseppe Michelucci's contextual post-war reconstruction adjacent to the Ponte Vecchio.

by Angiolo Mazzoni from 1938 to 1942, and the station was completed between 1947 and 1950 by teams working under Eugenio Montouri and Annibale Vitellozzi. In *Italy Builds,* it is the one contemporary exception to the series of contextless object buildings, and with regard to the archaeological appropriation of its major contextual feature, the American critic G.E. Kidder Smith gushed:

> This fourth-century-B.C. wall in the very midst of this startling modernity emphasizes as nothing else could to the traveller the roots and never-dying cultural contribution of the most fascinating city in the world.[7]

In the completed project, a great glazed hall communicated directly with the piazza outside. The sense of contemporaneity produced through this scheme was modified by two elements: the presence of a section of the Servian Wall which passed through the facade at an oblique angle and formed one boundary to the space (Figure 14.2); and the attachment of the new hall, its oversailing roof linking with the rather more monotonous forms of the station beyond, designed during the fascist period.[8] The asymmetrical vault of the main roof reflected the profile of the Servian Wall so that its silhouette could be appreciated by waiting passengers. The cantilevered roof and canopy were offset compositionally by the wall of office accommodation behind, with ribbons of continuous windows coursing across the travertine facade that acted as a modern screen to the arcuate forms of the station side buildings. The cross galleria, open

FIGURE 14.2 Stazione Termini, Rome, showing the relationship with the remains of the ancient Servian Wall.

at either end and connecting into the street network of the city, was identified as serving as a mid-twentieth-century equivalent to Milan's Galleria Vittorio Emanuele, if lacking that structure's enduring glamour. The station represented the new political situation of the country as an open, modern, democratic veneer placed over a substantial fascist legacy. Stazione Termini featured heavily in the cinema of the period, most notably in the eponymously titled 1953 film featuring the American stars Jennifer Jones and Montgomery Clift, their voices dubbed into Italian. But the significance of the structure as an emblem of the new social context saw it also provide a brief setting for a short sequence in the anthology *L' Amore in Citta* (1953), and also in Federico Fellini's *Nights of Cabiria* (1957), where some sense of the anonymity of the modern city was expressed in the reluctance with which the heroine meets a prospective suitor amidst its crowds.

The building of the Stazione Termini demonstrated that skill in the design of the public realm had not disappeared with fascism, and that new architectural languages could be used to create significant contemporary places. Many significant architectural figures who had been prominent under fascism, such as Luigi Moretti and the aged Marcello Piacentini, continued their careers, although their former political affiliations caused them some professional problems.[9] However, in this period two central architectural figures came to the fore: Bruno Zevi (1918–2000) and Ernesto Nathan Rogers (1909–69). Zevi's championing of organic architecture perpetuated battles with academicism that were already half a century old, but his attempt to recast architecture promoted the appreciation of space as fundamentally an architectonic rather than an urban phenomenon. The internal logic of a building was seen to have primacy over its external appearance; indeed, in Zevi's book *Architecture as Space: How to Look at Architecture,* the Vittorio Emanuele Monument in Rome was specifically condemned because it had no internal rationale and consisted exclusively of external rhetorical motifs.[10] The implicit political message was one of complete rejection of what the monument represented, indeed of the representational qualities of architecture per se. In contrast, an architecture which grew organically from its interior uses was proposed, with the urban form left as the organic by-product of internal architecture. The logic of this route was that buildings should express individual identity rather than any reciprocity, and in application, the resulting void between buildings would simply fall victim to the motor traffic which was beginning to dominate the urban scene, as private car ownership proliferated with the 'economic miracle'.

By contrast, Rogers' position was more dependent on creating a continuity of experience of design, in his phrase, 'from the spoon to the city', where rational processes of organization and manufacture would be harnessed to fulfil social needs.[11] Rogers, who was the principal CIAM (Congrès Internationaux d'Architecture Moderne) correspondent in Italy, held a nuanced position, with the editorship of the journal *Casabella* providing the opportunity to promote a reflective debate about architecture and the city. He was prepared to acknowledge Italian rationalism's debt to fascism as a focus of youthful idealism in a form of *apologia.* In the introduction to *Italy Builds* he remarked on how the congruence of Italy's experience under fascism and the development of modernism in architecture lent it a unique perspective.[12] The language of the International Style had already been formulated in Germany and France before the young Italian architects, the *Gruppo Sette,* announced their adherence to its ideals in 1927.[13] The enforced adoption of conventional materials, due to the international embargo on steel during the late 1930s, had also given Italian modernists the spur to explore traditional forms as a means to extend the language of the new architecture. This particular history offered post-war architects a distance from the standard international language of

FIGURE 14.3 Torre Velasca, Milan, viewed from the roof of the duomo.

modernism, the new context against which such experiences were to be deployed being rampant commercial redevelopment, often of a hostile nature, with little regard for a devalued historic environment now seen as a locus for mass tourism. In contrast, the professional and academic re-evaluation of traditional urban form for its typological value, which emerged as a concept following the war, became allied with an economically pragmatic concentration on context and a scientific approach to the analysis of urban forms. In Milan, Rogers would transform the work of his practice, BBPR, from the cool abstraction of the open cube memorial to the deportees in the Cimitero Monumentale (1945–55) to the super-medievalism of the Torre Velasca (1958) (Figure 14.3). The latter structure adopted an expressive form in its provision of new office and residential accommodation, its energies displayed in its dramatically jettied silhouette suggesting that this leading modernist practice had adopted a conservative stance in relation to urbanism.[14] Indeed, the circle of younger architects around this milieu (such as Aldo Rossi who will be discussed later) were, like their contemporaries in Team 10, keen to expose the failings of contemporary urbanism as experienced in the new developments.

Writings on urban space and architecture

While Italian architects were concerned with internal debates and with the many social problems of their country, it was left to others to reassess and identify the qualities and unique characteristics of Italian architecture and urban space. In contrast to the internal debates, this external scrutiny came initially in the form of a series of publications which attempted to analyse both the historical and the contemporary situation, and which sought to bring some definition and clarity to a confused situation. The first publication was the article 'The Third Sack' by Henry Hope Reed, in *The Architectural Review.* On the pretext of informing the numbers of pilgrims who would be drawn to Rome for the Holy Year of 1950, Reed placed recent developments against the historical context of the Renaissance and the baroque city. The rearrangement of Piazza Venezia for the creation of the Monument to Vittorio Emanuele received particularly unfavourable coverage, as an example of a significant historical space which had been turned into a hostile stream of traffic.[15] The second publication was *Italy Builds: Its Modern Architecture and Native Inheritance* by Kidder Smith, a survey of contemporary Italian architecture published in 1955, again showing quite a marked division between the historic and the modern. The book's two sections, dealing with contextual matters and recent building respectively, reinforced the implied lack of continuity between past and present. The traditional images which were referred to as exemplary subjects were explicitly urban, but in contrast the contemporary examples cited were (with the one exception already noted) isolated buildings, with the urban context absent, or at least cropped out of the photograph. Published bilingually in English and Italian, the implicit message of this book was that the new forms presented in the second half of the book, despite their use of modern materials and non-traditional forms, were as dependent on the territorial and climatic context of the Italian peninsula as the historical examples in the first section, although the connections were hard to discern.[16]

The last publication, which did not appear until 1963, was the product of a specifically British townscape sensibility which as well as surveying the grand manner of urban composition was adept at identifying the secondary order of signage, decoration, street furniture and daily usage which vivified an urban environment. It was the *Architectural Review*'s publisher H. de Cronin Hastings who produced *The Italian Townscape* under the pseudonym Ivor de Wolfe. It was a compendium of photographs, descriptions and diagrams that sought to account for the phenomenon in a comprehensive manner.[17] But in this instance townscape was essentially content to concern itself with the optimistic sun-filled images which appealed to aficionados from colder climates, the surface dressing of urban space rather than its substance, its gaze firmly averted from the huge new developments which characterized Italian urbanization in this period and which Italian architects and film-makers were exploring.

The revaluation of urban space

Fellini's *Nights of Cabiria* (1957) emphasized the internal exile of the periphery dwellers, never more so than when Cabiria, in a scene cut from the original release, witnessed charitable support given to people reduced to living in caves. True to the spirit of Neo-Realism, the film was shot on actual locations, where the typical activity depicted – prostitution and the developing Roman night life – could be recreated for the film with some authenticity. Disaffection with the limits of reality, however, would determine Fellini's later work, although few viewers of *La Dolce Vita* (1960) initially realized that it had been shot on a set, the economically vibrant city of spectacle apparently being replicable in cinematic form. The cultural validation of such a simulation on film therefore prefigured the imitation of traditional urban space in architectural practice.

La Dolce Vita presented a new Italy, one which had abandoned its traditional values and was intent on hedonism, although its urban spaces acted as a form of code for social audacity. While more bourgeois and materially comfortable than those depicted in Neo-Realist cinema, the peripheries of the city are still presented as bleak, in contrast to the glamour of Rome's central spaces, and in particular the allure of the Trevi fountain, where Anita Ekberg seduced the hero Marcello. Public spaces were reappropriated after the self-denial of the post-war years. The economic miracle which Italy was witnessing had renewed confidence in public display, both at the level of the family, with the growth of car ownership, and at that of the nation, with Italy hosting the 1960 Olympic Games – the fascist context of most of the sporting venues in Rome being politely overlooked. But in *La Dolce Vita* the re-envisioned historic public space was similarly shorn of any direct political meaning, and any signification which might refer to unpalatable political circumstances was ignored, and replaced with the aesthetics of sensation. Spaces which had previously been intended to evoke historical memory and the endurance of various distinctive values were now to be regarded as backdrops for the spectacular moment, operating as disposable and interchangeable elements, sometimes in cinematic techniques artificially replicated or simulated by back projection.[18] The availability of such sites – especially in the major centres of Rome, Florence and Venice – to increasing numbers of tourists undermined the sense that they might be authentic manifestations of Italy's urban culture. Prosperity brought in its train disillusion with the status quo and the desire to break through the beautiful surface image presented by Italian cities to the world.

Observation of the Italian scene attracted those who claimed to see beyond appearance. The international visitors who came to the modernist CIAM conferences in Bergamo in 1949 and Venice in 1950 were presented with a country where the projects being produced by the younger generation of architects were firmly within orthodox modernist models, even if they did make some acknowledgement of vernacular precedent in their use of materials. However, the universal models favoured by the older generation of modernists, as well as the intuitive approach of townscape, would both be rejected by Team 10, for whom *the* great European archetype was the Italian hilltown, free of association with planned propaganda, and redolent of an urban life which emphasized community rather than order. The work of Giancarlo De Carlo (1919–2005) in producing the masterplan for Urbino between 1958 and 1964 was the most successful demonstration of this approach and initiated a long series of architectural interventions in the city and its outskirts.[19]

De Carlo's role was pivotal, despite his relative isolation; he was a figure well connected in the leading international architecture networks through both Team 10 and his earlier connection to Rogers, with whom he had collaborated in CIAM. Yet the uniqueness of his position

FIGURE 14.4 Piazza Maggiore, Bologna, with the Facciata dei Banchi.

in these arenas was matched by his detached position in Italian politico-architectural culture, since (if one accepts the paradox) his affiliation to anarchism meant that he had a distanced attitude to the aesthetics of the traditional city. On the one hand, his study of Urbino was an exemplary demonstration of the validity of urban conservation policies during an era of wholesale redevelopment. But, on the other hand, the university buildings introduced into the fabric of the city, most notably the *Magistero,* concealed their novel architectural forms behind relatively anonymous exteriors, as if afraid to be seen intruding on the form of the Renaissance city, abdicating the provision of public appearance to the historical past, a gesture that might be regarded as radical only in its modesty.[20]

Beyond his interest in conservation, De Carlo's architectural and urban intentions would fail to be communicated, because his highly individualistic work stood apart from that of others in Team 10, which was by turns banal or mechanistic. The attempt by this circle to abstract the rules of social patterns from traditional architectural forms, while simultaneously denying themselves the comfort of those forms and their characteristic materials, undermined the faith in a modern urbanism from within the architectural profession at exactly the same time as the utopian project of modernism was under attack from without. However, the aesthetic attention lavished on the Italian urban environment in pursuit of an ultimately elusive humanistic modernism would provide material with which a younger generation of architects, both Italian and foreign, could reconstruct the idea of the European urban tradition and, within it, the significance of the legibly public space.

Policies which foresaw a use for traditional urban forms had a surprising origin in approaches pursued in communist-controlled Bologna.[21] Ideologically, the city council was keen to frustrate commercial development, as a means of protecting the interests of the working-class quarters which occupied the city centre. These mutually supportive motives could therefore be used to create an image of the city which was based firmly on its past, but which would come to be regarded and internationally acclaimed as startlingly progressive. The values first successfully implemented in Bologna would then be adopted by other major Italian cities, which had already experienced some degree of destructive improvement. The conservation of historic centres had been codified in the Gubbio Charter of 1960, and that document was to influence the study commissioned from Leonardo Benevolo by Bologna's city council in 1962–65. This study brought together the typological studies developed by Saverio Muratori with proposals for how the essentially medieval city form could be helped to function more efficiently in the very changed circumstances of the later twentieth century. Functional and dimensional limits thereby encouraged the maintenance of residential quarters in historic city centres, preserving the signs of daily life which would prove attractive to that most unscientific of markets, cultural tourism. The pedestrianization of Piazza Maggiore in Bologna in 1968 (Figure 14.4) was presented as a political signal in favour of equality (against the prioritization of private car ownership), the appropriation of what would come to be seen as 'green' urban policies and the validation of the aesthetics of the traditional urban environment. With the widespread political unrest developing in 1968, and the slowdown in construction in the wake of the energy crisis of 1973, the urban stage was set for the values demonstrated in Bologna to be exported to other European centres.[22] The social value of defined urban space, indeed its value as a component of the architectural project, would return to prominence and would contrast with the discredited work of the recent past. In this scenario, the appearance of another avant-garde, Superstudio, a group of architects based in Florence in the late 1960s, brought together two particular aspects of the urban culture of that period. Firstly, there was the critique of consumerism represented by the Italian post-war consensus and the favouring of radical solutions to social and political problems. Secondly, the vulnerability of the historic urban environment, revealed by the flooding of Florence and Venice in 1966, suggested the possibility of liberation from the constraints of architectural and urban conservation.[23] Instead, Superstudio proposed an environment which dispensed with notions of property division, of good design as commodification, and above all with the traditional restrictions of place. Superstudio thereby established a distinct Florentine architectural culture, less historically and politically driven than the schools of Rome, Milan and Venice. Despite the beginnings of political unrest and violence, this was a time of optimism when the possibilities of economic growth and social change seemed limitless.

What distinguished Superstudio from their contemporaries was the comprehensiveness of their vision. Although always keen to provoke, they adopted an architectural language which they applied to the design of objects, buildings and cities. Exemplified by their project 'histograms', their Cartesian forms managed to suggest both the rationalism of contemporary corporate forms and the global mysticism of the counter-culture. A good deal of professional skill was required to span this division, through an easily identifiable production which had learnt the graphic lessons of Pop Art. As one of their collaborators commented:

> Living in a city no longer means inhabiting a fixed place or urban street, but rather adopting a certain mode of behaviour, comprising language, clothing and both printed and electronically transmitted information; the city stretches as far as the reach of these media.[24]

Rossi, Pasolini and the poetics of the city

Against this optimistic utopian prospect, the energy crisis and subsequent economic downturn of the early 1970s would bring to the fore a more equivocal vision of Italian urbanity. Emerging from the search for architectural and urban form which favoured the typological approach (referred to variously as *la Tendenza* or Neo-Rationalism), Aldo Rossi (1931–97) was first a significant urban theorist, and then the creator of potent architectural images in drawn and built form.[25] For Rossi, the political context of this focus was the type of authenticity which intellectuals often ascribed to working-class life, rather than a romanticization of previous social conditions, but from his early career his work was to be suggestive of nostalgia. However, in a period marked by considerable social unrest, Rossi's position on the political Left led to him being banned from teaching in Italian universities by the government, and hence to his pursuit of an academic career outside Italy.

The continuous process of urban history was a theme developed in Rossi's *The Architecture of the City*.[26] Published in Italian in 1966 and translated into English in 1982, the book presented a tough critique of the modernist city but used Marxism to argue for an almost fatalistic adherence to the *zeitgeist*. Rossi proposed that architecture stood outside the fluid tide of history, dependent for its power on the qualities of its geometry and its accumulation of patina from surviving over time. This placed great emphasis on the collective experience of the city, and consequently reduced the individualizing tendencies of the unique monument, giving this analysis a significant focus on the issue of permanence in architectural typology. Rossi's text distinguished surface appearance from context, the atmosphere of a city being apparently replicable without any comprehension of the typology from which it was built, although this division would be a phenomenon which would bedevil the future broader reception of his own work. Mistrustful of the proponents of contextualism on grounds of their subjectivity, he developed a brand of rationalism that posited an architecture and urbanism which was less apologetic about its presence, but which acknowledged that time would transform it, that time would domesticate the urban intervention through use. The ambiguity of this position in relation to the temporal dimension contrasts with the fixity with which contextualists appropriated a past point in history.[27]

Similar ambiguity marked the work of Pier Paolo Pasolini and infused his reading of the Roman townscape, particularly in his two early films *Accattone*[28] (1960) and *Mamma Roma* (1962). A prevailing Christian iconography was shown as lying beneath the surface of *borgate* low-life, most explicitly in the earlier film. In one of its initial scenes, the eponymous Accattone's dive into the Tiber from Bernini's Ponte Sant' Angelo is framed like a Renaissance 'Baptism of Christ'. This metaphor is reinforced in the finale, through the hero's death on the emblematically modern Ponte Testaccio in the presence of two thieves, accompanied on the soundtrack by a chorale from Bach's St Matthew Passion. This contextual passage from baroque Rome to the contemporary city, albeit a modernity peopled by faces which could have stepped from Caravaggio's paintings of the early *seicento,* occurred again in the later film, with Mamma Roma defined as a profane Mater Dolorosa, eventually exiled to a housing development (designed in reality by the typo-morphologist Muratori). Pasolini, as John David Rhodes has observed, used the prosaic actuality familiar from Neo-Realism in an overtly poetic form, paralleled by the typology of the banal which characterized the writing and practice of Rossi.[29] Pasolini's presentation of the city is postmodern in its eclecticism and historical reference, and his film work would increasingly follow an interest in fables which left the problems of realist context behind. In a short contribution to the anthology *Amore e Rabbia* (1969),

Sequenza del fiore di carta, the actual location of Via Nazionale – itself historically significant as the first modern street in Rome and politically suspect because of its commercial and governmental character – is the scene of a tracking shot where the typically Pasolinian male character progresses in a carefree way, intercut and collaged over with images from recent wars, until we see his body lying bleeding in the vicinity of the Monument to Vittorio Emanuele at Piazza Venezia. To Marxists like Pasolini or Rossi, the authenticity of such an architectural location was just a veil over the political situation of oppression and struggle which would inevitably be played out in urban space. In a similar manner to Pasolini, the poetic content of *The Architecture of the City* broke out from the quasi-documentary tone it adopted as a cover. Influenced by the left-leaning utilitarianism of contemporary architectural theory, it moved beyond the common explanation of vernacular typology to the then unfashionable study of monuments. Characteristic of its origins during the high period of International Style modernism, Rossi's book concentrated on buildings rather than spaces. However, one passage dealt with the Roman Forum and used that space as an example which fulfilled many themes: the relationship between topography and place, between the planned and the organic, between history and the present. His reference to its timeless qualities is explicit in not depending on the counter-intuitive meaning of separation by history, but rather on the contemporaneity of its historical existence. He considered the life of the Forum in the following terms.

> People passed by without having any specific purpose, without doing anything: it was like the modern city, where the man in the crowd, the idler, participates in the mechanism of the city without knowing it, sharing only in its image. The Roman Forum was thus an urban artefact of extraordinary modernity; in it was everything that is inexpressible in the modern city.[30]

For Rossi, the Forum as an urban artefact embodied the essence of the city of Rome, displaying the palimpsest of its history, its point of origin and its contemporary centre. It is a space and context which could not be replicated elsewhere, only evoked through analogy. Its history represented the type of metamorphosis that Rossi implied would be the desired fate of his own buildings, their regularity and geometry eroded by habitation and use, but accommodating that change and persisting through time, accreting memory.

Within this intellectual milieu, Rossi's early substantial *projects* approached the issue of urban space in a stealthy manner, as if breaking cover would impede the success of his strategy to recover urban values in architecture. The housing Rossi designed in 1969–70 at Gallaratese on the outskirts of Milan revealed the particular characteristics of his evocative use of typology. The principal public feature is the portico which runs the length of the block (Figure 14.5), on two related levels, the junction of which is negotiated by a monumental set of steps and four over-scaled cylindrical columns.[31] The regularity of its form reflected its origins in traditional types of Lombard housing, but its refusal to articulate the uses to which its public element might be put meant that it was regarded as heartlessly oppressive and interpreted as a late flowering of fascism. Rossi's principal references, historical tradition and the experience of the modern, were shared with fascist architecture. But he was working in a context where historical form had been mistrusted and modernity had become an internalized search for novelty. Rossi had no embarrassment about the context in which he had been formed, and he was concerned to step outside the futile search for the contemporary, in the manner of his great hero, the early-twentieth-century Viennese architect Adolf Loos.

FIGURE 14.5 Gallaratese housing, Milan: Aldo Rossi's use of public space in the form of a colonnade beneath the apartments.

The poetic invocation of the peripheral city also features in Rossi's most celebrated project, the New Cemetery of San Cataldo in Modena, designed between 1971 and 1978 and still incomplete.[32] The phased construction did not begin until the early 1980s, but the extensive publication it attracted has ensured that the haunting images of the drawings and photographs captured the quintessence of Rossi's work.[33] His critical attitude to the relationship between past and present was revealed, as he evoked the work of his own immediate predecessors and recent history. In an echo of the open cube of BBPR's memorial to the deportees in Milan, at the centre of the Modena project is the sanctuary, a hollow cube open to the sky and punctured with regularly spaced square openings (Figure 14.6). It has the appearance of an abandoned construction site, of an incomplete project such as featured in many Neo-Realist films. Its presence represented the transitory nature of contemporary life in a fugue on the collective housing unit.[34] The abandoned house and the deserted factory are among the urban images which Rossi exploited for the elegiac qualities they evoke, yet there are other meanings and echoes, indicative of a broader historical perspective. Although they feature colonnades similar to that at the Gallaratese housing, the barracks-like quality of the *columbaria* of the New Cemetery distils the memory of the barracks of the Fossoli camp, built outside the nearby town of Carpi, the deportation point for all the victims of the Nazi occupation of Northern Italy. With his sensitivity to the resonances of banal forms and their echo of the horrors of the twentieth century, the ambiguities of Rossi's aesthetic choices are apparent. They express a context which not only was related to a common culture but, at close proximity to such a significant site,

FIGURE 14.6 The New Cemetery of San Cataldo, Modena. Aldo Rossi's cubic sanctuary occupies the central space of the projected complex.

was also local and historical. The ultimate collective experience was exploited to inform the contemporary city about the loss of urban space, and a nostalgic atmosphere appeared to have replaced faith in technological and social progress.

Although misinterpreted because of its superficial similarity to fascist architecture, Rossi's deliberate use of barely resolved junctions between concrete, stucco and steel held poetic echoes of the industrial landscape, which was redefining Italian urban form during these years. Their rhetoric was based on the common left-leaning technique of the heroization of the working life and environment, a utopia of the ordinary, which Rossi had witnessed in the Soviet Union. His recovery of past forms was not dependent on amnesia with regard to the modern city, but the assimilation of its divergent strands. Rossi's modest stance was that the city was beyond the capacity of design as control, that its political status had a symbiotic relationship with its form, where ends and means became one. And yet for the viewer of Italian films, as much as for the observer of Italian cities, the distraction provided by the fragment of a facade, the balustrade of a bridge or an embankment, the skyline of a city, continued to evoke a specific vision of urban life which raised it beyond the ordinary. As Pasolini wrote in his poem 'The Tears of the Excavator',

> . . . Stupendous, miserable,
> city, you made me
> experience that unknown
> life, you made me discover
> what the world was for everyone.

Acknowledgement

Material from this paper is adapted from Canniffe (2008) *The Politics of the Piazza: The History and Meaning of the Italian Square,* Aldershot: Ashgate.

Notes

1. Shiel, M. (2006) *Italian Neorealism: Rebuilding the Cinematic City,* London and New York: Wallflower.
2. Tafuri, M. (1989) *History of Italian Architecture 1944–1985,* Cambridge MA and London: MIT Press, pp. 3–48.
3. Cheles, L. (2001) 'Picture Battles in the Piazza' in Cheles, L. and Sponza, L. (eds) *The Art of Persuasion: Political Communication in Italy from 1945 to the 1990s,* Manchester: Manchester University Press, pp. 124–79.
4. Rowe, P.G. (1997) *Civic Realism,* Cambridge MA: MIT Press, pp. 100–116.
5. Seavitt, C. (1998) 'Cinecitta', *Dimensions: Journal of the College of Architecture and Urban Planning at the University of Michigan,* Vol. 2, Ann Arbor: University of Michigan, pp. 129–31.
6. Di Biagi, F. (2003) *Il Cinema a Roma: Guida alla storia e ai luoghi del cinema nella capitale,* Rome: Palombi Editori.
7. Kidder Smith, G.E. (1955) *Italy Builds: Its Modern Architecture and Native Inheritance,* London: Architectural Press, p. 232.
8. Ferrero, M. (2004) *Architetture di Pietra nella Roma del Novocento,* Rome: Palombi Editori, pp. 94–104.
9. Kirk, T. (2005) *The Architecture of Modern Italy Volume 2: Visions of Utopia 1900–Present,* New York: Princeton Architectural Press, pp. 149–53.
10. Zevi, B. (1957/1974) *Architecture as Space: How to Look at Architecture,* New York: Horizon Press, p. 33.
11. Branzi, A. (1984) *The Hot House: Italian New Wave Design,* London: Thames and Hudson, p. 8.
12. Kidder Smith, *Italy Builds,* pp. 9–14.
13. Etlin, R. (1991) *Modernism in Italian Architecture 1890–1940,* Cambridge MA and London: MIT Press, pp. 225–38.
14. Kirk, *The Architecture of Modern Italy Volume 2,* pp. 170–74.
15. Reed, H.H. (1950) 'Rome: The Third Sack', *The Architectural Review,* 107 (638) February: 91–110.
16. Kidder Smith, *Italy Builds.*
17. De Wolfe, I. (1963) *The Italian Townscape,* London: Architectural Press.
18. Bass, D. (1997) 'Insiders and Outsiders: Latent Urban Thinking in Movies of Modern Rome' in Penz, F. and Thomas, M. (eds) *Cinema & Architecture: Méliès, Mallet-Stevens, Multimedia,* London: British Film Institute, pp. 84–99.
19. De Carlo, G. (1970) *Urbino: The History of a City and Plans for its Development,* Cambridge MA: MIT Press.
20. Zucchi, B. (1992) *Giancarlo De Carlo,* Oxford: Butterworth; Blundell Jones, P. and Canniffe, E. (2007) *Modern Architecture through Case Studies 1945–1990,* Oxford: Elsevier/Architectural Press, pp. 165–76.
21. Benedetti, S. (1973) 'For a Typological Hypothesis in the Restoration of Historic Centers' in Bardeschi, M.D. et al. (eds) *Italian Architecture 1965–1970,* Florence: IsMEO, pp. 164–71.
22. De Pieri, F. and Scrivano, P. (2004) 'The Revitalization of Historical Bologna' in Wagenaar, C. (ed.) *Happy: Cities and Public Happiness in Post-War Europe,* Rotterdam: NAi, pp. 452–59.
23. Lang, P. and Menking, W. (eds) (2003) *Superstudio: Life without Objects,* Milan: Skira.
24. Branzi, *The Hot House,* pp. 63–66.
25. Kirk, *The Architecture of Modern Italy Volume 2,* pp. 182–85.
26. Rossi, A. (1982) *The Architecture of the City,* Cambridge MA and London: MIT Press.
27. Ibid.
28. Italian for 'beggar', this is the nickname of the film's main character, Vittorio.
29. Rhodes, J.D. (2007) *Stupendous, Miserable City: Pasolini's Rome,* Minneapolis and London: University of Minnesota Press.
30. Rossi, *The Architecture of the City,* p. 120.

31. Moschini, F. (ed.) (1979) *Aldo Rossi Projects and Drawings 1962–79,* London: Academy Editions, pp. 52–57.
32. Blundell Jones and Canniffe, *Modern Architecture through Case Studies,* pp. 189–200.
33. Barbieri, U. and Ferlenga A. (1987) *Aldo Rossi Architect,* Milan: Electa, pp. 46–53.
34. Eisenman, P. (1979) 'The House of the Dead as the City of Survival' in Frampton, K. (ed.) *Aldo Rossi in America 1976–79,* New York: Institute for Architecture and Urban Studies, pp. 9–15.

INDEX